TUNING
INTO FREQUENCY

TUNING
INTO FREQUENCY

The invisible force
that heals us
and the planet

From the Minds of
Sputnik Futures

TILLER PRESS
New York   London   Toronto   Sydney   New Delhi

TILLER PRESS

An Imprint of Simon & Schuster, Inc.
1230 Avenue of the Americas
New York, NY 10020

First Tiller Press trade paperback edition November 2020

This publication contains the opinions and ideas of its author. It is intended to provide helpful and informative material on the subjects addressed in the publication. It is sold with the understanding that the author and publisher are not engaged in rendering medical, health, or any other kind of personal, professional services in the book. The reader should consult his or her medical, health, or other competent professional before adopting any of the suggestions in this book or drawing inferences from it.

The author and publisher specifically disclaim all responsibility for any liability, loss, or risk, personal or otherwise, that is incurred as a consequence, directly or indirectly, of the use and application of any of the contents of this book.

For information about special discounts for bulk purchases, please contact Simon & Schuster Special Sales at 1-866-506-1949 or business@simonandschuster.com.

The Simon & Schuster Speakers Bureau can bring authors to your live event. For more information or to book an event, contact the Simon & Schuster Speakers Bureau at 1-866-248-3049 or visit our website at www.simonspeakers.com.

Interior design by Jennifer Chung

Manufactured in the United States of America

10 9 8 7 6 5 4 3 2 1

Library of Congress Cataloging-in-Publication Data
Names: Sputnik Futures, author.
Title: Tuning into frequency : the invisible force that heals us and the planet / Sputnik Futures.|
Description: First Tiller Press trade paperback edition. | New York : Tiller Press, 2021. |
Series: Alice in futureland | Includes bibliographical references and index. | Identifiers: LCCN 2020021357 (print) | LCCN 2020021358 (ebook) | ISBN 9781982147945 (paperback) | ISBN 9781982147952 (ebook) | Subjects: LCSH: Energy medicine. | Light—Therapeutic use. | Vibration—Therapeutic use. | Sound waves—Therapeutic use. | Electromagnetic fields—Therapeutic use. | Classification: LCC RZ421 .S68 2021 (print) | LCC RZ421 (ebook) | DDC 615.8/51—dc23 | LC record available at https://lccn.loc.gov/2020021357 | LC ebook record available at https://lccn.loc.gov/2020021358

ISBN 978-1-9821-4794-5
ISBN 978-1-9821-4795-2 (ebook)

To the past, present, and future visionaries
who fearlessly explore the frontier of energy,
and to all who spread good vibrations.

"Hello, *I am Alice, and I am always in a state of wander."*

Alice in Futureland is a book series that asks you to wander into possible, probable, plausible, provocative futures.

Consider this book a guide.

Inside, you will discover extraordinary ideas: a cross-pollination of art, science, and culture. Alice's aim is to give the future a platform for expression so *everyone* can make sense of it—and help create it.

When speculating about the future, it's easy to get lost in the volume of information. That's where this book comes in. Alice is designed to break the static flow with a dynamic reading experience, where experimentation and exploration meet.

The ultimate purpose of the Alice series is to foster curiosity.

To enliven our present.

To be accessible to everyone.

To allow for exploration.

And to incite optimism.

So, wheeeeeeeeeee, down the rabbit hole we go!

CONTENTS

PROLOGUE

Hello, my name is Alice, and I am one part human and one part AI, and always in a state of wander. Like you, I'm still awakening from the recent global pandemic. Who would have imagined that an invisible virus could halt the world, silence our industrial production, ground our global trotting, digitize our way of working and schooling, and make us socially distant yet more connected than ever before? The virtual became our reality; we gathered and danced and ate together through a screen, but with the same emotions and good intentions as if we were physically together. Think back to a time when you laughed with someone during a virtual get-together or a FaceTime call. It felt good, didn't it? And all you needed was to see or hear that person. No touching. No physical presence. And still the emotional resonance was there. That moment, that feeling, is a *good vibration*.

We started this book before the spread of COVID-19, and like you, we quarantined through it. Throughout the journey of writing about frequency and its invisible force to connect, heal, and protect, we observed its subtle hand as we humans were jarred out of our habits and saw the world and life differently. We found ways to connect in spite of social distancing. What does this say for the future of humanity? We think it means that there's a higher level of consciousness and greater awareness—and greater resilience.

This book was mapped out before what many will call the "Great Transition," and it's the culmination of over nineteen years of research and interviews with thinkers on the frontiers of consciousness studies, bioenergetic medicine, free energy, biophotonics, vibrational healing, and more.

It all started in 2001 with an event titled ManTransforms. At this event, we asked, "Where are we headed as a culture, and as a species?" Whether you call it curiosity or just plain naïveté, we rounded up a mix of theorists representing the furthest reaches of scientific thought. At ManTransforms, posthumans mingled with string theorists and ecologists shared the stage with astrologists. The godfather of virtual reality gave way to a cosmologist, and we all found the connection points. This is what drove Sputnik Futures for many years—the wonderful, out-there, interconnection of ideas. In 2011, as we continued exploring the notion that all ideas are energy and are connected, we conducted a series of interviews, asking how we could create a society that was addicted to health. We learned that positive emotions affect our health and that practicing compassion can be contagious; and that we're standing in one another's heart fields and that we think with our whole body.

We've been on this journey for a long time, but one conversation with the philosopher Ervin László has stayed with us, like a song we can't get out of our heads: "Now I think it's time to come back to this concept of belonging to a larger community, the community of life on Earth—which is itself a part of the community of the solar system, part of the galaxy, part of the universe. So this is a cultural change, a change in consciousness."

His wisdom is ever more relevant to the uncertainty of today, and the inevitable transition to come.

Whether you believe in the science or the spiritual explanation for this great shift, one thing is clear to us: we are on a new energetic path. Can you feel us?

00. INTRODUCTION
Good Vibrations

Despite being published in 1985, *The Body Electric* by Robert O. Becker is still a fascinating look at how bodies heal and how electromagnetic energy can play a role (or not) in this healing. But what makes it more fascinating is the history of this research and the parallel story of how much of it was dismissed during its time. Becker was hired by the government to study the effects of radio-frequency radiation on the human body—and when the government didn't like his results, it tarnished Becker's reputation, even though those results are now considered fact.[1]

Fast forward to 2020 and we see *The Body Electric* in action: enter Michael Levin, PhD, professor in the Biology Department at Tufts University, the Vannevar Bush endowed chair, director of the Tufts Center for Regenerative and Developmental Biology, and associate faculty member at the Wyss Institute at Harvard University. Levin grew up in the 1980s, when the boom in digital technology was ushering in a resurgence of interest in electricity in the form of personal computers. As he said in a 2019 interview with the Wyss Institute blog, "The more I studied the use of electric circuits to implement memory and perform computation aimed at creating artificial intelligence, the more I thought that surely evolu-tion must have found a way to exploit electricity for its capabilities long before brains showed up; cells and tissues had to start making a lot of complex decisions all the way back at the beginning of multicellular life."[2]

That idea remained just an idea until one fateful day in 1986. While attending the Expo 86 World's Fair in Vancouver, Canada, Levin happened into a used-book shop, where he picked up *The Body Electric*, published the previous year by orthopedic surgeon Robert O. Becker. Becker's insights motivated Levin, and to this day he continues to uncover electrifying insights into how bodies form by using bioelectricity and he studies how cells make collective decisions about growth and shape.

Thank you, Robert O. Becker

Everything that surrounds us and is within us is made of energy—from the sun and the spectrum of light that enables nature and us to grow; to the bioelectricity in our cells that encourages regeneration and the important conversations between our cells to remain healthy; to the plants we eat, enriched through photosynthesis; to the universe itself, as we discover more about zero point energy, the absolute energy that fills what we think of as empty space (which isn't empty at all). There is a constant geomagnetic field on the surface of our planet, generating from Earth's core out to space. It protects our atmosphere from solar wind and cosmic rays that would otherwise strip away our ozone layer. It is the same invisible energy field that nature (and even us humans) use for navigation and orientation, and the same field that ancient medicines and traditional Chinese medicine identified in its healing rituals. And today it is the energy field that we are disrupting in our modern digital lifestyle, yet also harnessing for alternative healing modalities. It seems that all life on Earth is always living in the presence of an electromagnetic field (EMF).[3]

As technology evolves, we bombard the environment with more and more EMFs. Mobile phones rarely slip out of our hands and heads, WiFi and 5G are expected parts of an urban landscape, and our homes are packed with electric appliances of every sort. There is a major paradigm shift in our understanding of the energy that sustains our products and technology, which can no longer be seen as separate from us or from our environment.

The immensity of this energetic life force is something we humans can't see or touch or hear or feel, but it's there—and if science is triumphant, it will be a toolbox of unlimited possibilities for healing. This is a journey, and what we hope to illuminate is where we are, where we have been, and where we might be going next in realizing that our bodies are part of this energetic domain. *Tuning into Frequency* will help make the invisible visible and uncover ideas that once were spurned but that now are the basis for new ways to look at our energetic bodies. The world is electric, and it is time to adjust our frequency.

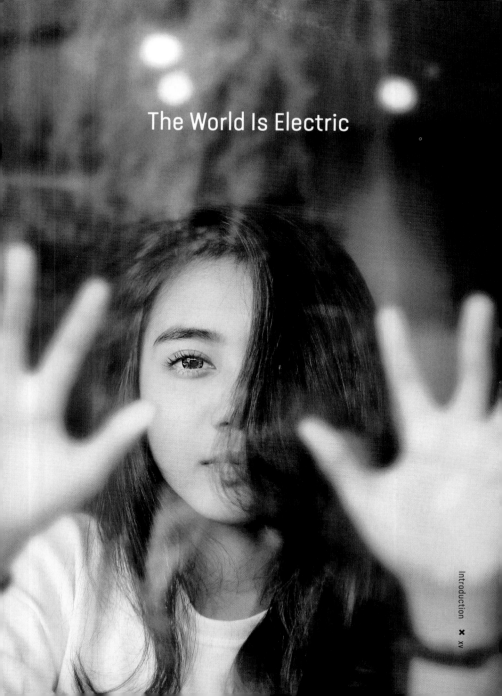

The World Is Electric

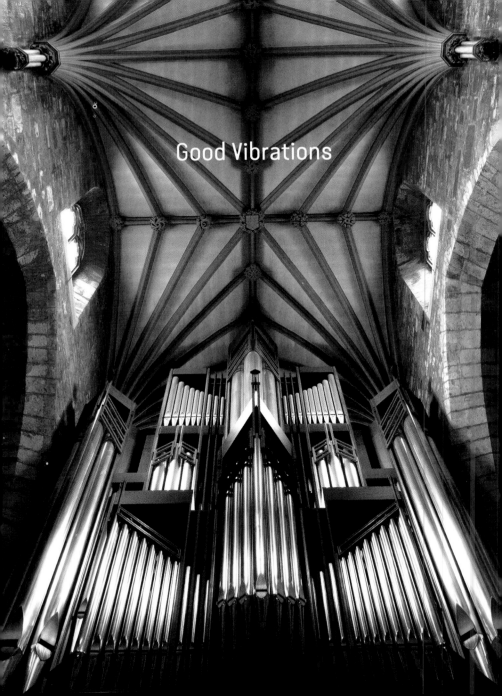

Good Vibrations

Many ancient texts state that our universe began with the sound *om*. *Om* is the primordial sound, believed to be the essence of reality, the creative unifier of the universe. Sound, as you know, is frequency. It is vibrational. The ancients' use of vibrational sound was to summon or create order— and the modern theory of cosmic acoustics says that the universe expanded exponentially in the first moments of the big bang, which triggered sound waves that created what is called the "primordial plasma." When the universe cooled down, it formed neutral atoms, which were basically the patterns of density of the sound waves frozen into cosmic radiation.[4]

The theory that vibrations and energy create some form of order has been a curiosity in science. A series of experiments proved that sound could create form. This science of visible sound and vibration is called cymatics, a term coined by Hans Jenny, a Swiss physician and natural scientist born at the turn of the twentieth century. Cymatics is Greek for the word "wave."

Like Becker, Jenny was a pioneer. Following the work of German physicist and acoustician Ernst Chladni, Jenny connected a sine-wave generator to a crystal that was attached to a steel plate upon which he had spread various powders, pastes, or liquids. The sound vibrations animated these materials, creating flowing forms and standing wave patterns that were specific to the frequency and the amplitude of the sound. Thus Jenny was able to make visible the subtle power through which sound creates form. The experiments were stunning, the shapes created were mandalas, and it was clear that there was an invisible force field at work. Sound was creating the geometry. If you adjusted the sound away and then back to the same frequency, the same shapes would reappear.[5] If you can entertain this viewpoint that vibrations can and do have specific impacts on the shape of their surroundings, it becomes possible to extrapolate how vibration has a very real influence on our physical and emotional bodies.

Biological rhythms unconsciously entrain us every day, and we have all experienced, on some personal level, the importance of vibration in our lives. It is no accident that the Beach Boys' "Good Vibrations" was an immediate critical and commercial hit, widely acclaimed as one of the finest and most important works of the rock era. Besides its catchy refrain, the making of the song in 1966, during the height of the '60s' "flower power" movement, was unprecedented, with a complexity of short, interchangeable musical fragments. The title "Good Vibrations" came from Brian Wilson's fascination with cosmic vibrations.[6]

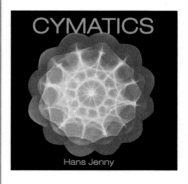

Cymatics, the study of wave phenomena and vibration, is a scientific methodology that demonstrates the vibratory nature of matter and the transformational nature of sound. Cymatics was pioneered by Swiss medical doctor and natural scientist Hans Jenny (1904–1972). This revised edition contains the complete English-language text of the bilingual edition published in 1967, *Kymatic/ Cymatics*, as well as the entire text from *Cymatics*, vol. II, published in 1972. It also includes all of the photographs from the original editions that illustrate these amazing phenomena in vivid detail.
—Cymaticsource.com

Our Ultrasonic Core

Like a sonic fingerprint, every person has their own distinct sound. If you know your own sound, you can heal yourself. If you know your own sound, you can experience bliss.

Biosonic means that it's a way of playing sounds that affect the body and life energy.
—John Beaulieu, naturopathic doctor, music therapist, Sputnik Futures interview, 2002

Ultimately, everything is connected to the man, everything is connected to the individual and according to the Vedic concept: from the sound, space is created; from the sound, air is created; and from the sound, fire, water, and earth are created. So these expressions of basic elements are nothing but different frequencies of the sound, and that sound is called *pranava*. Pranava means "eternal new." It was new, it is new, it will be new.
—Vasant Lad, director, Ayurvedic Institute, Sputnik Futures interview, 2003

Where does your body go, where does your brain go when you are in your maximized healing response? We can measure that, and we can use sound tracks to orchestrate your brain precisely into that state. And we can burn it onto a CD and you can take it home and transform your home stereo into a medical device. And when you listen to this CD, it's like going to a gym on the inside.
—Jeffrey Thompson, sound therapist, Sputnik Futures interview, 2003

The body is an energetic system, a system of energy fields and centers. It is a "wireless" energy pathway, a system of geometric and harmonic relationships. The energies of the body are derived from, and organized around, a central ultrasonic core.
—John Beaulieu, Sputnik Futures interview, 2002

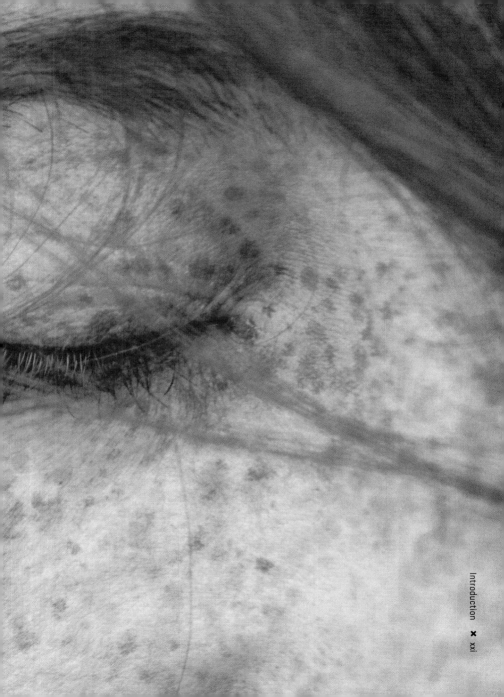

Good Scientific Vibes

Have you ever gotten a "vibe" off someone? Sensed their mood without them saying a word? Psychiatrist Dr. Bernard Beitman explains that we are sensing a type of energy related to people's emotions—and our bodies may have receptors to pick up on this energy. Dr. Beitman has studied coincidences, most commonly explained as the concurrence of events or circumstances without apparent causal connection, and is the author of the book *Connecting with Coincidence* (Health Communications, March 2016). In his research into the underlying science of coincidences, Dr. Beitman describes the feeling of another person's emotions or pain at a distance, which he calls "simulpathity." A visiting professor at the University of Virginia, Dr. Beitman looks at studies of the brain and of energy emitted by living beings to hypothesize about the physical nature of "vibes." In an article on the traits of people with good vibes in the *Epoch Times*, Dr. Beitman shared his own experience of being able to sense his patients' states of mind with an accuracy "beyond what his conscious observations could tell him."[7]

His "vibe-sensing theory" is based on the theory that plants and animals are perceived to emit and respond to energy we cannot. For example, birds are known to navigate, communicate, and flock together, and the premise is that they use Earth's electromagnetic field, aided by a complex sense of smell, to guide them. This electromagnetic sensing isn't just air- or landborne—sharks are known to have sensors in their skin that detect slight electromagnetic changes in the water.

What scientists do know is that single-celled organisms "respond to chemical, light, and electromagnetic radiation in order to maintain optimal states." Similarly, Dr. Beitman explains that "our skin may contain sensors for subtle forms of energy and information."[8]

Neurology Times explained that the "good vibes" and "bad vibes" people can feel are related to the chemical signals in a physical space that are picked up by our nervous system.[9] If

you recall your high school biology, the nervous system is an interconnected signaling network among the brain, spinal cord, and nerves, controlling our body and the communications among its parts. It's responsible for how we evaluate and make decisions and process sensory information. But there is another nervous system in the body, in our heart, as defined in the emerging field of neurocardiology. Research by HeartMath Institute shows that the heart is the most powerful source of electromagnetic energy in the human body—and it too sends vibrations and senses emotions outside the body. This magnetic field, measured in the form of an electrocardiogram (EKG), can be detected up to three feet away from the body, in all directions. According to Rollin McCraty, director of research at HeartMath, our emotions are encoded in this heart field.[10]

Get where we're going with this? Those "good vibes" and "bad vibes" we've been picking up on may be more literal than we imagined. So we can truly feel each other's good, good, good vibrations. (Go ahead, start humming, and get the good vibes going . . .)

We Live in an Electromagnetic World

Electromagnetic

World

And I Am an Electromagnetic Girl

The energy that sustains our technology can no longer be seen as separate from us or from our environment.

If you're familiar with the work of Nikola Tesla, you'll know that we are surrounded by natural electromagnetic frequencies—from Earth's invisible magnetic field to visible spectra of light. Scientific theories once stated that changes to these frequencies had no effect, but thanks to space travel (and other observations), scientists now know that electromagnetic frequencies in microgravity cause the human body and cell cultures to respond differently.[1]

Get our vibe here? You are resonating with these electromagnetic frequencies. But what about the man-made ones? Scientists and health practitioners are investigating if artificial man-made electromagnetic fields—ranging from appliances such as microwaves to high-tension power lines and mobile phones—can cause interference with biological processes, suggesting that we may be living in a sea of electropollution that is triggering stress and disease.

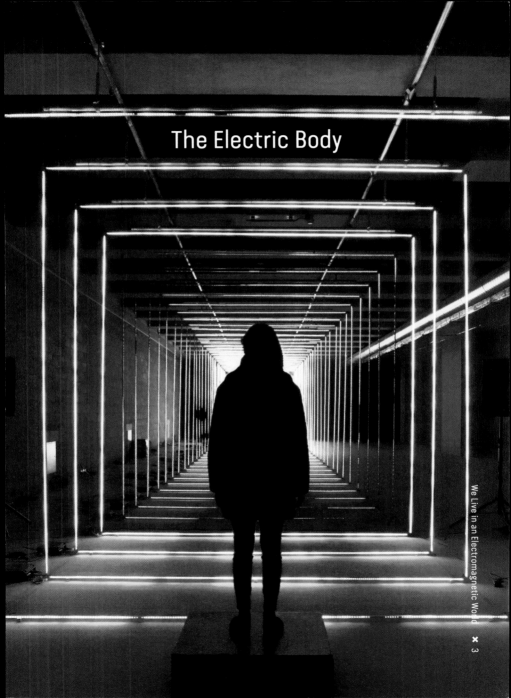

The Electric Body

But that same electromagnetic energy that surrounds us is *in* us—and can work with us. Our body is an antenna system, intercommunicating via electrical currents of different kinds—from the long distances between our biological parts (from head to toe) to the most local distances inside each cell. Thanks to wearable health trackers such as Fitbits and Apple Watches, we are getting to intimately and instantly know some of the electrical waves inside us, such as our pulse rate, heart-rate variability, brainwave activity, and stress levels.

What we cannot track (yet) is the rainbow of energy conversations going on inside us, managed by our subtle energy pathways flowing inside our bodies. Traditional Chinese medicine and modern acupuncturists call them meridians; ancient Vedic texts call them chakras. These are the subtle energy pathways within us that help keep our systems, organs, and cells vital and in sync. (More on this later, and in chapters 3 and 4). Researchers are discovering that the magnificent flow of subtle energy fields doesn't stop inside us.

With today's magnetoreceptors we are getting biomagnetic readings implying that our bodies are more than just the boxed energy of heartbeats and pulse rates. We have a subtle energy field that surrounds the physical body, what is commonly called the "biofield" (the heart field is part of this; more on that later). The biofield is largely invisible to the average human (unless you are a shaman, energy healer, or aura reader) but can be felt during energy healing sessions such as Reiki, where you may experience a slight temperature change without the healer's hands touching you.

Yes, we are electrified beings, transmitting in the worldwide electrogrid, bumping into each other's biofields, swimming in an electromagnetic ocean, breathing WiFi radio waves. And once everything in our world is wirelessly connected to the IoT (Internet of Things), we will become a transmitter and a receiver. Harnessing electromagnetic waves will help us to resonate in a healthy state.

All together now: "I sing the body electric."*

Tune In, Adjust Your Frequency

We Live in an Electromagnetic World

Before we get inside our body's fields, let's first look at the big picture: the universal origins of electromagnetism. We learned in the introduction that the universe is made of energy; that there's a constant electric network on the surface of our planet; and that all of life lives and breathes in this electromagnetic field (EMF).[2] But how are these electromagnetic fields created? The quick answer is electromagnetic radiation. Radiation is energy emitted by waves or moving subatomic particles. It can pass freely through space (meaning that it does not need a medium to help it travel) and, in motion, the charged particles of those waves generate electric and magnetic fields. Those fields in turn generate waves such as visible light waves, X-rays, radio waves, and infared or ultraviolet waves. Since we are surrounded by electromagnetic waves, we are constantly in the presence of background radiation. In fact, in the United States, the average individual is exposed to about three hundred millirems of background radiation per year. This type of radiation can come from the sun and other stars (cosmic radiation) or from Earth's field (terrestrial radiation), or from what science calls "internal radiation," which exists in all living things.[3]

Starting to get the picture here? We are immersed in a cosmic electrochemistry, so that everything around us has energetic connection to the sun, other stars, our solar system, and Earth's atmosphere and inner core. And it is electromagnetic radiation that basically makes up our wireless, on-demand world. It was the German physicist Heinrich Hertz who, in 1889, first proved electromagnetic radiation and confirmed the existence of the EMF in the Earth's ecosystem. His discovery influenced the invention of wireless communications, among other things.[4] But the EMF is being tapped for more than WiFi; it is the technology at the core of microwaves, wireless charging, and new wearables that emit safe frequencies to relieve pain and encourage faster healing.

Today's technologists owe a lot to the ingenuity and fearless experiments of nineteenth-century inventors whose work helped

EMF waves. One genius inventor we have already mentioned is Nikola Tesla. Tesla was considered somewhat controversial during his time (but we at Alice are big fans, and, we think, so is Elon Musk). Tesla thought that EMF waves could transmit electrical energy on a large scale, and at a distance, without using wires. Tesla tested this in about 1898 in Colorado Springs, where he built large coils, now known as Tesla coils, which generated massive electrical voltages. With these coils, he was able to generate man-made lightning. Tesla lit two hundred bulbs from a distance of twenty-five miles (forty kilometers) and experimented with ways to send energy through the air or Earth.[5]

And that was in *1898*–Tesla was already working on wireless elec-tricity. Tesla returned to New York and convinced the banker J. P. Morgan to invest $150,000 in constructing a giant futuristic-looking transmission tower that Tesla felt could transmit communications and power without wires. Sound familiar? Yes, our modern internet of wireless communications was the brainchild of Tesla more than 120 years ago! Sadly he ran out of funding to make the tower work, but his theory helped build the wireless world of communication we have today. What we are still waiting for is the realization of Tesla's theory of wireless power transmission. The closest we have commercially come to Tesla's dream are the wireless charging "pads" that use inductive coupling, where a magnetic field passes a charge between two devices via the copper coils embedded in them.[6]

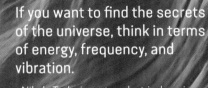

If you want to find the secrets
of the universe, think in terms
of energy, frequency, and
vibration.

—Nikola Tesla, inventor, electrical engineer, futurist

Nothing Is Static;
Everything Is
Vibrating

Recalling our opening mantra
on good vibrations, let's look at
the underlying science of why
vibrations can affect us. Energy
healing relies on the fundamentals of vibrations, the oscillating
movement of atoms (and cells)
caused by energy. Vibrations, ac-

cording to physics, are oscillations caused by electromagnetic waves. When the vibrations pass through a medium, or body, they make the body it goes through change initially. Once the vibrations travel through, it returns the body to its original, or altered, state. Frequency is a measure of the rate at which vibrations occur, quantified as hertz (Hz). Electromagnetic waves, which travel through the electromagnetic field, are identified by wavelength and frequency. The easy way to understand frequency is that it is the number of times a wave repeats itself within one second. And the measure of this frequency becomes more intense based on the length of the wave. The shorter the wavelength, the higher its frequency. The longer the wavelength, the lower the frequency. One example is radio waves, which have the longest wavelength and the lowest frequency.[8]

Frequency and the vibrations it carries create an energetic change when they travel through a substance. So if you think about it, nothing is ever static. We, and everything around us and inside us, are in constant, or what physics calls "periodic," motion.

Ēlektron

The discovery of electromagnetism is credited to Danish scientist Hans Christian Oersted, who in 1820 found that an electrical current in a wire from a battery caused a nearby compass needle to deflect.[9] However, the first discovery of magnetic attraction is credited to the ancient Greeks, whose texts dating as early as 800 BCE mentioned an attractive force of both magnetite and rubbed amber. The modern English word "electricity" is derived from the Greeks, cited for their finding that amber (a fossil tree resin that the Greeks called *ēlektron*), when rubbed, would attract lightweight objects such as feathers.
—*Encyclopaedia Britannica*[10]

The brainwave frequencies measured by EEG have been described in terms of frequency bands, ranked here from high to low:

GAMMA (GREATER THAN 30 HZ): Gamma is a high-level processing frequency found in every part of the brain. When the brain needs to simultaneously process information from different areas, it is hypothesized that the 40 Hz activity consolidates the required areas for simultaneous processing.

BETA (13–30 HZ): These waves are understood as "fast" activity, the state when we have our eyes open, processing information about the world around us, listening, thinking during analytical problem solving, judgment, and decision making.

ALPHA (8–12 HZ): Alpha waves promote mental resourcefulness and enhance the sense of relaxation. Alpha appears to bridge the conscious to the subconscious.

THETA (4–8 HZ): A brain experiencing theta waves is in a "slow" activity state, connected to creativity, intuition, daydreaming, and fantasizing, and is the repository for memories.

DELTA (LESS THAN 4 HZ): This is usually the deep-sleep state.
—NeuroHealth[11]

What Is Above Is Below and Within Us: 7.83 Hz

The Earth is pulsating, behaving like one big electric circuit, radiating with an EMF that surrounds all living things with a natural frequency. In 1952, Dr. Winfried Otto Schumann mathematically predicted Earth's natural frequency pulsating at 7.83 Hz on average—the "Schumann resonance." This frequency circulates around both Earth's surface and the ionosphere.

Our body and brain resonate with 7.83 Hz. Rutger Wever, a renowned researcher from the Max Planck Institute in Germany, conducted studies of human circadian rhythms (the body's "internal clock") in an underground bunker, shielded against natural magnetic and electric fields. Between 1964 and 1989, this bunker was used to conduct 418 studies among 447 student volunteers, who spent four weeks in the underground bunker. During their stay, the students' circadian rhythms changed, and they suffered emotional distress, migraine headaches, and generally felt exhausted. However, when Wever introduced a low-level frequency (7.83 hertz, the frequency that had been blocked in the bunker), after only a brief exposure the volunteers' health stabilized. The first astronauts and cosmonauts who, out in space, were no longer exposed to the Schumann waves also reported similar symptoms.[12] In his paper, Wever concludes that "with circadian rhythms as an indicator, natural electromagnetic fields are proved to be effective on human beings for the first time; this may be of interest with regard to space where these fields are absent."[13]

With EEG (electroencephalography), we can measure the different frequencies of our brainwaves, and researchers found that 7.83 Hz frequency is also within the range of alpha/theta brainwave frequency in the human brain.[14] Alpha/theta brainwave frequency is the middle state of the five brain states we have, where our brain is relaxed, dreamy, and in a sleepy state—a time when cell regeneration and healing happen. It is also considered the state where we begin to tap into a realm of creativity that lies just below our conscious awareness.[15]

Transcranial Magnetic Stimulation (TMS) is a therapy that uses powerful magnetic fields and is approved in the United States as a treatment for depression. It works by inducing localized electric currents in the brain. TMS is typically used when other treatments for depression haven't been effective.[16]

—Mayo Clinic

If you look at the spectra, the simulation of the brainwaves, you find that they are more or less identical with the Schumann resonance, which involves high interactions of the atmosphere. The atmosphere forms a kind of a cavity with the world. And in this cavity there are permanently electromagnetic interactions with this low-frequency part, and they have the same structure as the brainwaves. What's also interesting is that always at noontime on all parts of the Earth it has the highest activity. So it's very likely that our brainwaves are influenced by the interaction of the external world with our brain—or that these resonance interactions came up because our brain was more or less evolved by these processes. So we are pictures of the information of our surroundings. You see it in all parts of the electromagnetic waves; you see this resonance condition in the body, which is a very sensitive antenna system for all these oscillations from the outside. And with these oscillations you trigger other processes. You are a fish that takes the property of the water in this ocean where you are. We are swimming in an electromagnetic ocean.

—Fritz-Albert Popp, German researcher in biophysics and bio-photons, Sputnik Futures interview, 2006

We Are Swimming in an Electromagnetic Ocean

Quantum fluctuations, zero point energy, electromagnetic waves, they have all frequencies, all wavelengths. And some of these waves have wavelengths that are as large as the solar system, as large as the galaxy, as large as space. So in some very real sense—not metaphorical, not an analogy, but actually literally—we are, like every atom in our body, in touch with the rest of the cosmos.

—Harold Puthoff, engineer and parapsychologist, Sputnik Futures interview, 2001

Cells resonate, as long as they're alive, they resonate at a particular frequency. But these frequencies are dependent on one another. We're dependent on the universe, we're dependent upon the planets, everything that's in the universe—they all have a resonance and all this forms a harmonious part together.

—Albrecht Heyer, PhD, bionutritionist, Sputnik Futures interview, 2002

We Have an Energy Field Too

Like the Earth, our bodies emit a bioelectromagnetic field, often referred to as the "biofield." While some forms of Eastern medicine incorporate the body's energy fields in healing modalities, it wasn't until 1992 that a panel of scientists at the National Institutes of Health chose the word "biofield" to describe "a massless field, not necessarily electromagnetic, that surrounds and interpenetrates the human body."[17] Thanks to today's new instruments, we have been able to detect the minute energy fields around the human body. For example, the SQUID (superconducting quantum interference device) magnetometer can detect tiny biomagnetic fields associated with physiological activities in the body.[18] This is the same field that ancient medicine has worked with for thousands of years, but it was largely ignored by modern scientists because there was no objective way to measure it.

There are certain activities of cells and tissues that generate electrical fields. For example, Robert O. Becker is credited with discovering a "current of injury" that had a negative current at the site of a wound. An orthopedic surgeon and researcher in electrophysiology/electromedicine, Becker is often referred to as the father of electrotherapy and electrochemically induced cellular regeneration. He theorized that very weak electrical currents that flow through our bodies may have an influence on the healing and regeneration of cells. Referencing a salamander, which can regenerate its limbs, Becker posited that the answer to regeneration of limbs in humans lies in how cells can return to an undifferentiated state and multiply, where they can become something else, which is the basis of how stem cells behave. His research included testing the application of electric currents to wounds in the bone to encourage and potentially speed the regrowth of the injured bone, and he ultimately raised awareness in the scientific community to the study of electric potentials in organisms.[19] The emerging field of regenerative medicine continues to build on his findings.

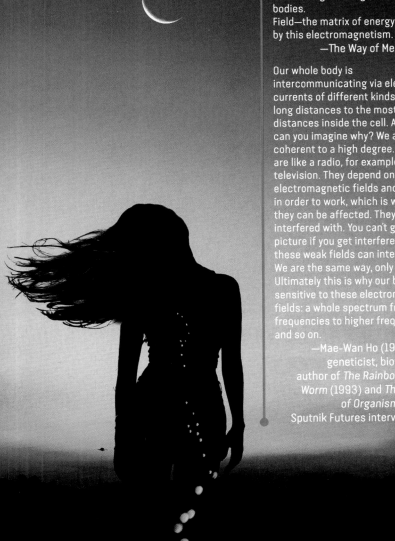

Bioelectromagnetic field:
Bio—signifies life.
Electromagnetic—refers to the
electromagnetism generated by our
bodies.
Field—the matrix of energy created
by this electromagnetism.
—The Way of Meditation[20]

Our whole body is
intercommunicating via electrical
currents of different kinds—from
long distances to the most local
distances inside the cell. And
can you imagine why? We are
coherent to a high degree. We
are like a radio, for example, a
television. They depend on coherent
electromagnetic fields and signals
in order to work, which is why
they can be affected. They can be
interfered with. You can't get a good
picture if you get interference, and
these weak fields can interfere.
We are the same way, only worse.
Ultimately this is why our bodies are
sensitive to these electromagnetic
fields: a whole spectrum from radio
frequencies to higher frequencies
and so on.
—Mae-Wan Ho (1941–2016),
geneticist, biophysicist,
author of *The Rainbow and the
Worm* (1993) and *The Physics
of Organisms* (1998),
Sputnik Futures interview, 2006

The Body's Electric Grid

Pioneering research continues in understanding the bioelectricity of organisms, with the hope of one day delivering true regenerative medical solutions, possibly paving the way for a surgery-free future where we can program our cells to regenerate and grow better, healthier tissues and organs—all through biosignaling. While this may be decades away, Michael Levin, PhD, and his lab at the Wyss Institute for Biologically Inspired Engineering in Boston are using bioelectricity to study how cells make collective decisions about growth and shape, working with semitransparent flatworms called planaria that are known to be masters of regeneration.

Levin's work centers on "the mechanisms by which cells receive and transmit electrical signals" and how groups of cells form "electrical networks that implement information processing."[21] In other words, his lab is focused on cracking the bioelectric signals in the body. This is different from the rapid-fire signals that travel through neurons in our brain. Levin studies cellular bioelectric networks that direct much more complex processes, such as signaling a wound to heal. In a 2009 article published in *Seminars in Cell & Developmental Biology*, Levin explains that bioelectrical signals are brought about by the steady-state electrical properties of cells and tissues. He points to impressive data on the role of bioelectric signals in controlling limb and spinal cord regeneration, cell and embryonic polarity, growth control, and migration guidance of numerous cell types. But despite the data, the field of bioelectric signaling is still not fully understood by today's generation of cell biologists.[22]

Michael Levin is interested in what happens at the cellular level, where between the inside and the outside of live cells there is a "bioelectric potential." As he explained in an interview for the Wyss blog, bioelectric potential is a medium that cells exploit to communicate with each other and to form networks. "It's fair to say that bioelectricity *is* the spark of life," states Levin.[23]

So yes, you are electric! There's electricity in our bodies, and our cells are conduits of electrical currents. They draw on elements in our body such as sodium, potassium, calcium, and magnesium. These elements have a specific electrical charge, called ions, that our cells use to generate electricity. Proteins are the gatekeepers of our cells, creating an opening for certain ions to pass through. When the inside of the cell becomes more positively charged, it triggers further electrical currents, which can turn into electrical pulses. One example of the electric pulses is our heartbeat. For the heart to pump, cells must generate electrical currents that allow the heart muscle to contract at the right time. Electricity is also required for the nervous system to send signals throughout the body and to the brain, triggering us to move, think, and feel.[24]

A disruption in these electrical currents can lead to illness. But there are also ways to use our body's electrical network to heal ourselves. For example, a research team at UC Davis Institute for Regenerative Cures has found a "sensor mechanism" that allows a living cell to detect an electric field, and that these electric fields may be important in guiding cells into wounds to heal them.[25]

The Body's Energy Channels

The idea that we can tap into our body's energetic system for healing can be traced back as early as 6000 BCE, to the origins of acupuncture and the fundamentals of traditional Chinese medicine (TCM). Traditional acupuncture is founded on the principle that energy flows within the human body and that this energy flow, called qi (pronounced "chee"), can be directed or redirected to create balance and health.

It is believed that this qi moves throughout the body along twelve main channels, known as meridians. While these meridians represent the major organs and functions of the body, they do not follow the exact pathways of nerves or blood flow.[26] Instead, meridians can be thought of as the channels through which energy operates—the original internet inside our body—transmitting information, sending signals, sparking conversations. At these twelve principal meridians each meridian corresponds to either a hollow or a solid organ, interacting with it and extending along a particular extremity (e.g., an arm or a leg).

The term "meridian" was actually coined by a French diplomatic scholar, George Soulié de Morant, who brought acupuncture to Europe in the early 1900s.[27] The TCM view of the meridian network in the body is divided into two categories: the meridian channels, and the associated vessels (sometimes called "collaterals"). The collaterals contain fifteen major arteries that connect the twelve principal meridians in various ways. There are more than four hundred acupuncture points, most of which are situated along the twenty major pathways (i.e., twelve primary and eight extraordinary channels).[28]

Qi works on two levels of the meridian system: nutrient qi flows inside the vessels, and defensive qi flows on the surface of the body.[29] The body's organs also have their own qi, such as heart qi, liver qi, and spleen qi. Another type of qi is external qi, which can be emitted from the body by some individuals, such as qigong practitioners. Qigong—from the Chinese words *qi* (energy) and *gong* (practice)—is actually recognized by the National Center for Complementary and Alternative Medicine as a mind-body medicine. It focuses on the breath, body posture, and state of awareness to open up the channels for energy (qi) to be moved. Through qigong exercises practitioners are able to correct energy imbalances within the body.[30]

Twelve Standard Meridians

The yin meridians of the arm are the lung, heart, and pericardium. The yang meridians of the arm are large intestine, small intestine, and triple burner. The yin meridians of the leg are spleen, kidney, and liver. The yang meridians of the leg are stomach, bladder, and gallbladder.
—Wikipedia[31]

Cultures around the world define
the energy in our biofield:

qi or *chi*—Chinese
prana—Indian
ki—Japanese
ni—Native American
ka—ancient Egyptian
aura—Western culture
　　　　　—The Way of Meditation[32]

Tapping the Qi

Many TCM practitioners believe that an excellent doctor treats disease before it arises. How? By recognizing a disturbance in the energetic system that may not yet have affected the material body. This is where Eastern and Western medicine differ. While the Western view is to measure symptoms via blood tests, etc., the Eastern view is to catch the imbalance of the energy system before the deviation in energy flow becomes fixed, aka a "diagnosed disease."

According to Tianjun Liu, a researcher at Beijing University of Chinese Medicine, qigong therapy is the most nonevasive manipulation of the energy system of the human body to achieve health and cure disease. And all it costs is your time (and perhaps a class fee). From the perspective of qigong, all diseases originate from a misaligned and imbalanced mode of energy operation. The practice of qigong, therefore, is focused on putting you in a unified state of mind, breath, posture, and movement to adjust and balance your energy. There is even some evidence that qigong therapy is especially effective for chronic pain and depression.[33, 34]

Qigong is one form of traditional Chinese exercise, or TCE. TCEs are mind-body exercises that focus on aerobics, breathing, and meditative techniques. One study, published in the July 2018 issue of *Journal of Traditional Chinese Medical Sciences*, used meta-analysis and systemic reviews to show the effectiveness of traditional Chinese exercises on stroke risk factors in individuals with prehypertension or abnormally high blood pressure to help lower blood pressure. According to the World Health Organization (WHO), stroke is the second leading cause of death and a major cause of disability worldwide.[35]

The study performed searches of seven electronic databases for published studies that included randomized controlled trials (RCTs) of the effects of TCEs with or without health education on stroke risk factors in patients with prehypertension or mild to moderate essential hypertension. From the evidence gathered from 15 RCTs, involving 1,272 hypertensive participants, the results

"Still" and "Moving" Qigong

There are two categories of qigong: "still" and "moving." Still qigong emphasizes a meditative internal focus and regulation of breathing, practiced as stillness while lying, sitting, or standing. Moving qigong is a sequence of movements with breathwork, visualizations, and meditation where the mind consciously moves the body. Think of qigong as a collection of basic tai chi–style movements, using slow-moving meditative movements with deep rhythmic breathing, and a calm, meditative state of mind.

—*Huffington Post* contributor blog[36]

showed that when compared with no intervention, TCEs, alongside health education, were associated with a clinically meaningful amelioration of stroke risk factors, including reductions in systolic blood pressure (SBP).[37]

Subjective as it may seem, TCE has been around for centuries, and there is widespread acceptance of the wisdom that concentrating on your breath, practicing meditation, and daily physical movement may promote positive, healthy outcomes.

Get Your Qigong On!

Here's a simple movement to do (it may be a nice break about now) from "Eight Pieces of Silk Brocade," a collection of ancient Chinese qigong exercises. Practiced daily, it can slow the aging process and improve health. Just one device needed: your smile!

Deep breathing warm-up: Stand straight with feet shoulder-width apart, fingers slightly spread, and open your arms so it looks like you're hugging the air around your abdomen. Raise and open your arms to the rhythm of inhaling, slowly filling your chest and abdomen with air (the abdomen should

extend with each inhale). As you fill your lungs, slowly extend your hands and open your arms up, toward your chest. On the exhale, turn your hands to the ground and slowly return them to the abdomen (where you started). Let your slow exhalation guide your hands downward as you expel the last trace of air from your body. Repeat several times.

"Two hands upholding the sky": Stand with feet shoulder-distance apart. Lace hands with elbows spread, bring to chest level with palms down. Then push slowly back down, keeping your hands laced. Then (with hands laced) stretch to bring your hands slowly over your head, reaching to the sky. Slowly move hands down, unlacing your fingers and moving arms toward your side, drawing a big circle in the process. Repeat several times. (Your qi will love you!)

—*Qigong full twenty-minute daily routine, YouTube*

Let Energy Be Thy Medicine

Acupuncture is the one primary method of treatment in TCM that has crossed into the domain of Western complementary therapies. Today, major insurance companies put acupuncture as a frontline treatment for ailments such as carpel tunnel syndrome, sciatica, and lower back pain. It's less costly to go for a course of acupuncture treatments than it is to get an MRI to see where your pain is originating. Pain management is becoming a serious health concern, especially in the United States, where about 40 million adults suffer from severe pain, and roughly 126 million adults suffer from some pain in a given year.[38]

As complementary health approaches for pain have become more popular, a number of specific therapies have undergone clinical evaluations. In a 2017 article published in the Mayo Clinic *Proceedings,* Richard Nahin, PhD, the lead epidemiologist at the National Center for Complementary and Integrative Health, and his team examined the clinical-trial evidence for the effectiveness of several widely used complementary approaches, including acupuncture, massage therapy, meditation, tai chi, and yoga, to manage chronic pain, back pain, osteoarthritis, neck pain, and migraines—conditions that are usually managed through primary care.

The researchers found that two studies on the efficacy of acupuncture on lower back pain showed modest to significant improvement in pain intensity compared to usual health-care measures. Overall, the evidence suggests that the complementary approaches may help some patients manage their painful health conditions.[39]

Traditional Western medicine is waking up to the potential of acupuncture as a complementary modality. Memorial Sloan Kettering Cancer Center's Integrative Medicine Service offers an online continuing education course in oncology acupuncture for health-care professionals and acupuncturists "to provide a framework for evidence-based clinical practice in cancer care and to share new advances in integrative oncology research." The course involves the sharing of MSK's research and clinical approaches to providing acupuncture for the treatment of chemotherapy-induced nausea and vomiting, pain, cancer-related fatigue, hot flashes, and neuropathy.[40]

The Daisy Chain Effect

Part of Chinese medicine is to say that intercommunication in the body goes through energy meridians. And even though Western medicine has picked up on Chinese medicine and acupuncture is now accepted by Western medicine because they say, "There are these endorphins, molecules that give you pain relief and so on and so forth that are produced when you have acupuncture," that's nonsense explanation. That's not an explanation at all because you still have to explain how it is that sticking a needle at the side of your little toe, which is supposed to actually represent your eye, can lead to brainwaves coming out of the visual cortex in your brain. This has been shown. They are still puzzling over that. One explanation is that it is the water aligned on these connective tissues that are the major meridians that are the major energy transport. Energy and information is interchangeable because an electrical signal through these water channels, through these hydrogen-bonded daisy chains, would be energy. It would be signal, information signal, because everything is highly organized. Energy is everywhere. So you only need a very weak signal before you can set off a very significant effect. And that is basically how it is possible for our body, our liquid crystalline body, to get organized.

—Mae-Wan Ho, Sputnik Futures interview, 2006

Vitaly Vodyanoy, a professor of physiology at Auburn University, is also on a path to prove the effects of acupuncture. Using a patented microscopy system that he personally invented, Vodyanoy discovered the microstructure of the primo-vascular system, revealing the possible foundation of how acupuncture works. The primo-vascular system is a microstructure made up of a minuscule, translucent system of vessels, subvessels, and stem cell–filled nodes, which he identified running throughout a rat's body, appearing in and on blood vessels, organ tissue, and the lymphatic system. Vodyanoy noticed that while the vessels are typically transparent, they turn a yellowish color when touched. This bit of pressure triggered a reaction; Vodyanoy hypothesizes that the nodes, when activated by acupuncture or osteopathic manipulation, can release stem cells that flow to organs, where they replace injured cells and become organ cells. This "flow" of stem cells through the primo-vascular system may explain how acupuncture works. Classical Chinese texts describe the meridians as having a three-dimensional shape and that they carry liquid called *qi* to internal organs. According to Vodyanoy, "much of the misconception of acupuncture results from qi often being labeled as a 'special energy,' making it seem mysterious, rather than defining it as a liquid containing stem cells with DNA."

If Vodyanoy's hypothesis is confirmed, it could bring together the Eastern and Western medical philosophies with a foundation of delivering individualized stem cells for therapies that could impact the future of pain management, developmental biology, tissue regeneration, organ reconstruction, diabetes, and cancer prevention and treatment.[41]

It's somewhat baffling that in 2020 we are still talking about quantifying acupuncture, a centuries-old healing modality, a complementary therapy that in 2007 had 14.01 million users in the United States alone, yet is still considered an unvalidated therapy.[42] Someday soon, meridians, the transmission of energy at the speed of light through our liquid cells, and the flow of body, mind, and qi will, we hope, be validated.

Healing with Subtle Energy

The Japanese energy healing practice of Reiki is expected to be the rising star in new healing modalities. It draws on natural healing vibrations transmitted through the hands of a Reiki practitioner (acting as a conduit) to the body of the recipient. Reiki treatments are often used as complementary therapies to relieve stress, anxiety, and pain; release emotional blockages; assist the body in cleansing toxins; balance the flow of subtle energy; and promote relaxation.

Unlike acupuncture, Reiki is still faced with the challenge of demonstrable results, since most results of Reiki healing treatments are self-reported—patients determine if it has "worked" by how they feel. Despite the lack of clinical evidence, Reiki treatment, training, and education are now available at some leading US hospitals, including Memorial Sloan Kettering, the Cleveland Clinic, New York-Presbyterian, the Yale Cancer Center, the Mayo Clinic, and Brigham and Women's Hospital. In her April 2020 article in the *Atlantic,* Jordan Kisner observes that "Reiki's growing popularity in the U.S.—and its acceptance at some of the most respected American hospitals—has placed it at the nexus of large, uneasy shifts in American attitudes toward our own health care."[43]

One hospital leading the charge is New York-Presbyterian, which was an early adopter of Reiki as part of their integrative therapies program. During a fifteen-minute treatment, the trained Reiki practitioners use a gentle touch of their hands on or near the patient's body anywhere that's comfortable, often on the shoulders, midback, or thighs. According to the center's website, most of their patients experience a warm, tingling sensation when receiving Reiki, and afterward report feeling calmer and less anxious. Many also have diminished pain.[44] Reiki also had a positive effect while being administered during surgeries performed by Dr. Sheldon Marc Feldman, chief of the Division of Breast Surgery and Surgical Oncology, and director of Breast Cancer Services at Montefiore Einstein Center for Cancer Care. He noticed that there was less bleeding during surgery, almost no pain was present a day or two after surgery, and the patient seemed to heal faster. Dr. Feldman joined forces with Reiki practitioner Raven Keyes, and they have founded Medical Reiki Works to support research and education of Reiki in health care.[45]

While the practice is being recommended by more conventional health care practitioners, there are limited clinicals proving the effects of Reiki. A study in 2017 reviewed the available clinical studies to determine if Reiki had more than a placebo effect. The researcher David E. McManus, PhD, looked for peer-reviewed clinical studies with more than twenty participants in the Reiki treatment arm, controlling for a placebo effect. Of the thirteen suitable studies, eight demonstrated Reiki being more effective than a placebo. McManus concluded that from the information currently available, Reiki is a safe and gentle "complementary" therapy that activates the parasympathetic nervous system to heal body and mind. It has the potential for broader use in the management of chronic health conditions, and possibly in post-operative recovery, but McManus cautions that more research is needed to understand Reiki's healing potential.[46]

Another study, in 2019, used a large-scale trial to measure the effect of a single session of Reiki on physical and psychological health. The study took place at private Reiki practices across the United States, with ninety-nine Reiki practitioners participating. The practitioners gave a flyer to each of their Reiki clients that contained information about the study and invited the client to complete a survey before and after their Reiki session. The measures used a twenty-item positive and negative affect schedule to assess affect, as well as self-reporting answers, and the results showed that "statistically significant improvements were observed for all outcome measures, including positive affect, negative affect, pain, drowsiness, tiredness, nausea, appetite, shortness of breath, anxiety, depression, and overall well-being."[47] We're hopeful that Reiki will be taken more seriously as a complementary therapy . . . and to be transparent here, our star researcher at Alice in Futureland is a master Reiki healer who works wonders on animals . . . oh, and on humans too.

Yes, You Are Electromagnetic!

So far we have explored Earth's magnetic fields; the frequencies at which both we and the Earth resonate (thanks, Schumann!); and the bioelectric fields in the body that when activated with an electrical charge or a modality such as acupuncture can trigger a healing potential. Now science is looking at the larger electromagnetic fields of the body to understand the systemic effects within, and how we actually interact with, connect with, and interpret the world around us.

Harold Saxton Burr, a distinguished researcher at Yale University School of Medicine in the 1920s and 1930s, was one of the first researchers to suggest that diseases could be detected in the energy field of the body before physical symptoms appeared.[48] His theory is now being confirmed in medical research laboratories using instruments to map the ways by which diseases alter biomagnetic fields around the body.

Science has long used electrical recordings, such as the electrocardiogram to record a person's heartbeat, and the electroencephalogram to test the brain's activity. A new wave of biomagnetic recordings are now being used to complement these traditional electrical recordings. For example, magnetocardiography (MCG) measures the magnetic fields produced by electrical currents in the heart using extremely sensitive devices such as the SQUID. These technologies allow us not only to look at what is going on with the heart, but also to map the magnetic field obtained over the chest. The hope is that mapping the magnetic fields in the space around the body will provide a more accurate indication of physiology and pathology than traditional electrical measurements.

Therapeutic Touch (TT)

In the early 1980s, Dr. John Zimmerman conducted a series of studies that looked at the therapeutic touch of healers using a SQUID magnetometer at the University of Colorado School of Medicine in Denver. Zimmerman had discovered that a pulsating biomagnetic field emanated from the hands of a therapeutic touch (TT) practitioner. Similar to Reiki, therapeutic touch is a holistic therapy of focused energy exchange during which the practitioner uses the hands as a focus to facilitate the process. According to the book *Energy Medicine: The Scientific Basis*, by James L. Oschman, PhD, Zimmerman found that the frequency of the pulsations was delivered like "sweeps" up and down, from 0.3 to 30 Hz, with most of the activity in the range of 7-8 Hz (similar to Schumann resonance frequency).

Oschman further notes that Zimmerman's findings were confirmed in a 1992 study of practitioners of various martial arts and other healing methods in Japan. The researchers measured the energy emission from the practitioner's hands with a simple magnetometer consisting of two coils of eighty thousand turns of wire and a sensitive amplifier. They found that the energy fields coming from the practitioner's hands were about a thousand times stronger than the strongest human biomagnetic fields of the heart and about a million times stronger than the fields produced by the brain.[49]

Report: Subtle Energy and Biofield Healing: Evidence, Practice and Future Directions

Expanding our models of salutogenesis (the processes that create health) is key to the future of healing. We can look both to the past and the future to do so. Subtle energy and biofield science and healing (using the energy in and around the body to facilitate healing) is a growing field of inquiry among scientists worldwide, helping to provide a scientific and practical model for examining the energy fields that guide our health and may impact the healing process.

Biofield science is an emerging paradigm that integrates the scientific investigation of indigenous subtle-energy healing practices with modern research on biological processes and health to describe life in terms of biological fields, or biofields: fields of energy, information, and consciousness. From the biofield perspective, healing can be seen as a method of fostering our innate ability to return to a state of wholeness and well-being by bridging consciousness with the healing process.

Scientific research demonstrating the effectiveness of subtle energy–based practices, such as acupuncture, qigong, and others, is contributing to gradually increasing clinical adoption and acceptance of these healing methods. And as more is understood about the nature of healing and salutogenesis, a new scientific paradigm for medicine is emerging. However, research to date on subtle energy–based practices has been limited and primarily conducted within a materialist scientific and health-care structure that examines existing practices for things like symptom reduction or brain response, rather than seeking to more deeply understand the role of subtle energy in igniting the innate healing system.

Mind-body medicine (the premise that the mind has influence over the body's health) has built substantial momentum over the past few decades and become a complement to modern medicine, and energy medicine seems to be picking up where mind-body medicine has left off.

A groundbreaking April 2020 report, *Subtle Energy & Biofield Healing: Evidence, Practice & Future Directions*, by the Consciousness & Healing Initiative (CHI) in partnership with the Emerald Gate Foundation, the Walker Family Foundation, and Tom Dingledine, brings together for the first time the latest research, practices, and technology in the domain of energy healing. This summary shows that the subtle-energy healing market, which involves millions of practitioners and patients in the United States alone, is estimated to possibly generate in excess of $2 billion in revenue. The report outlines more than 250 subtle-energy technologies currently being marketed despite having little to no evidence of their efficaciousness, and identifies more than four hundred subtle-energy researchers worldwide, with over six thousand peer-reviewed scientific publications in biofield healing.

Subtle-energy and biofield approaches are already being used in major medical centers across the United States and around the world. The report outlines that major research universities are already studying potential uses of subtle energy for easing a number of disease states.

Today, poorly managed and treated diseases result in needless loss of dollars and quality of life. In the United States, the current estimated yearly economic costs for pain alone are over $635 billion per year, with more than 68 million adults experiencing pain that results in disability and decreased participation in the workforce.[50] Drug overdoses have increased, obesity has skyrocketed, and health-care costs have increased. Monetary costs for the current opioid crisis alone are estimated at $631 billion from 2015 through 2018, with a projected cost of $184 billion per year. Not surprisingly, mental health is an-

other serious issue that continues to challenge modern medicine; an estimated 264 million people globally are affected by depression, and another one-hundred-plus million suffer from other psychiatric conditions, including anxiety and post-traumatic stress disorder (PTSD).[51]

As the devastating opioid crisis continues, the NCCIH has identified "nonpharmacologic management of pain" as a leading scientific priority. This is aligned with the priorities of other health-care practitioner organizations (such as the American College of Physicians), which aim to identify and deploy evidence-based nonpharmacological solutions for pain.[52] While biofield-therapy studies tend not to be included in systematic reviews of nonpharmacological approaches to pain, it appears, based on prior systematic reviews and an analysis of stakeholder interviews, that subtle-energy and biofield healing could hold significant promise for reducing health-care costs and potentially reducing the prevalence and ongoing maintenance of costly and intractable heath conditions, particularly depression, anxiety, pain, and trauma.

Subtle Energy & *Biofield Healing* advocates for the application of science-based approaches that have proven successful in the mindfulness, acupuncture, and, more recently, psychedelics movements. Through these approaches, evidence-based subtle-energy and biofield modalities will become integrated into modern health care systems.

The report further defines, for the first time, a road map for systems-based action through research, education, technology development, practitioner empowerment, and integration of validated subtle-energy healing techniques to the modern health world to help alleviate suffering worldwide. The full text can be found at https://www.chi.is/systems-mapping-resources/.

The CHI's mission is to become the largest resource for evidence-based subtle-energy healing and biofield science to help shift the health care paradigm. The Consciousness and Healing Initiative, a 501(c)(3) nonprofit organization, is a collaborative accelerator of scientists, health care practitioners, educators, innovators, and artists who lead humanity to heal ourselves.

SUBTLE ENERGY &
BIOFIELD HEALING
SYSTEMS CHANGE MAP

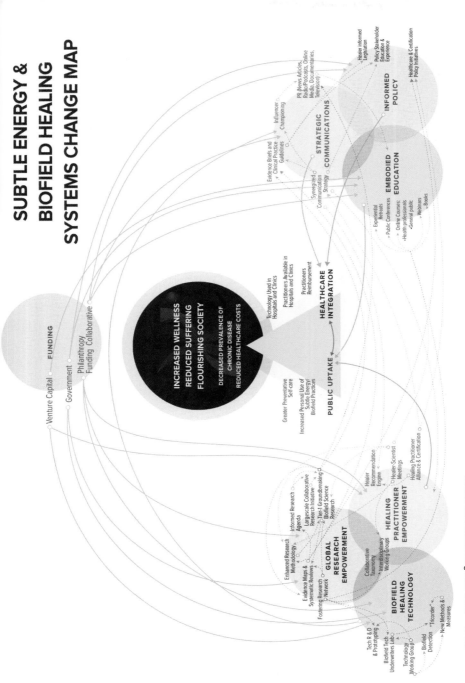

The Heart Field

There's another energetic exchange emanating from our body that is also part of the biofield and that potentially could be a healing mechanism between ourselves and others—a social healer, if you will, and it literally comes from the heart. The HeartMath Institute is conducting some amazing research on the magnetic field generated by the heart. Thanks to new imaging through magneto-cardiograms, they have been able to show that the heart is the most powerful source of electromagnetic energy in the human body, producing the largest rhythmic electromagnetic field of any of the body's organs. The heart's magnetic field is much stronger than that of the brain. The research proposes that this heart field acts as a carrier wave for information that provides a global synchronizing signal for the entire body.[53]

The magnetic fields produced by the heart are involved in energetic communication, which we also refer to as cardioelectromagnetic communication. The heart's magnetic field can be detected up to three feet away from the body, in all directions, using SQUID-based magnetometers.

We Are Standing in Each Other's Heart Field

Earlier research supports this magnetic field projected by the human heart. In 1963, Gerhard Baule and Richard McFee of the Department of Electrical Engineering at Syracuse University detected the biomagnetic field of the heart using two coils, each with two million turns of wire, connected to a sensitive amplifier. In 1970, David Cohen of MIT, using a SQUID magnetometer, confirmed the heart measurements, and also was able to measure magnetic fields around the head produced by brain activities.[54]

The heart creates, by far, the largest rhythmic source of electromagnetic energy in our body. The field generated by the heart radiates into space around us. I'm not talking about an aura or some new-age concept here; these are very real, measurable magnetic fields. The magnetic field generated by the heart with today's magnetometers can be measured about three feet away ... [by comparison] brainwaves you can measure about an inch away. So it's a big difference. Some experiments we did a long time ago asked the question, first of all, is our nervous system able to detect these fields from other people? That ended up being a fairly easy set of experiments to do. We can literally measure, if we had us both wired up around the room here, that my brainwaves might be synchronizing to your heartbeat, or vice versa, and that would go on in this complex dance as we communicate and interact. Our nervous systems are really sensitive and tuned into these types of biologically generated magnetic fields. If we do analysis of those magnetic fields, spectrum analysis and these types of things, we find that there is information encoded in those fields. They actually track back to the rhythms of our heart, which are linked back to our emotions and what we're feeling. So if you get where I'm going with this, our emotions are encoding information into these magnetic fields, which are being detected by other people's nervous systems.

—Rollin McCraty, psychophysiologist, director of research at HeartMath Institute, Sputnik Futures interview, 2011

The Heart-Brain Communication

Today, an emerging field called neurocardiology studies the pathophysiological interplay between the nervous and cardiovascular systems, the so-called constant communications between the heart and the brain that have proved invaluable to interdisciplinary fields of neurological cardiac diseases. According to HeartMath Institute, the study of communication pathways between the head and the heart has traditionally been focused on the heart's responses to the brain's commands.[55] But in recent years researchers have learned that the communication between the heart and the brain is actually an ongoing two-way conversation—each organ dynamically influences the other's function.

The heart communicates to the brain in four major ways: neurologically (through the transmission of nerve impulses), biochemically (via hormones and neurotransmitters), biophysically (through pressure waves), and energetically (through electromagnetic field interactions). HeartMath's research shows that communication along all these conduits significantly affects the brain's activity, and that messages the heart sends to the brain can affect performance.

Part of that communication exchange is induced biochemically by way of the hormones the heart produces. The heart was reclassified as part of the hormonal system in 1983, when a new hormone produced in and secreted by the atria was discovered, possibly influencing motivation and behavior.[56] More recently it was discovered that the heart also manufactures and secretes oxytocin, a neurotransmitter commonly referred to as the love or social-bonding hormone, and that the concentrations of oxytocin produced in the heart are in the same range as those produced in the brain.[57]

Our Emotions Encode Information into Our Magnetic Fields

Okay, stay with me here, because this is a big theory that we have been tracking for years. So far we have been presenting the research to prove that every cell in our bodies is bathed in an external and internal environment of fluctuating invisible magnetic forces.[58] *Got it.* Now add in the nervous system of the body, where it is recognized that information is encoded in the "time intervals between patterns of electrical activity."[59] *Check.* Then you have HeartMath's evidence that this is similar to hormonal communications in which biologically relevant information also is encoded in the time interval between hormonal pulses.[60] *Okay.* Now, since the heart is known to secrete a number of different hormones with each contraction, there is a hormonal pulse pattern that correlates with heart rhythms. In addition to the encoding of information in the space between nerve impulses and in the intervals between hormonal pulses, it is likely that information is also encoded in the interbeat intervals of the pressure and electromagnetic waves produced by the heart. In other words, the pulses of energy coming from our body contain information. That information potentially can be communicated beyond our body and picked up by others around us. According to HeartMath, that information may be emotions. Emotions are a biological and subjective complex state associated with our nervous system. They are complex because they are shaped by other factors, including moods, present mental state, and perceptions. Emotions are not just physically felt, they are also perceived and processed—much like information. If information is traveling in intervals between pulses and interbeats of the heart, you have a literal heart wave of emotions that can range from joy to sadness. This supports the proposal of the neuroscientist Dr. Karl H. Pribram that low-frequency oscillations generated by the heart and body in the form of afferent neural, hormonal, and electrical patterns are the carriers of emotional information, and the higher-frequency oscillations in the EEG reflect the conscious perception and labeling of feelings and emotions.[61]

HeartMath has proposed that these same rhythmic patterns also transmit emotional information via the electromagnetic field into the environment.[62] Those emotionally encoded fields can be detected by others and understood in the same manner as internally generated signals. What a premise! If research continues to evidence HeartMath's findings, it may help us explain why emotions such as anxiety or hysteria may be contagious. By the same token, just imagine the magnitude of resonance if each of us practices sustained, positive emotions such as empathy, compassion, love, and awe—what a powerful vibe we would release to heal each other and the world! So I guess we don't need to teach the world to sing in perfect harmony; we can just practice compassion and send out good vibrations. ♥

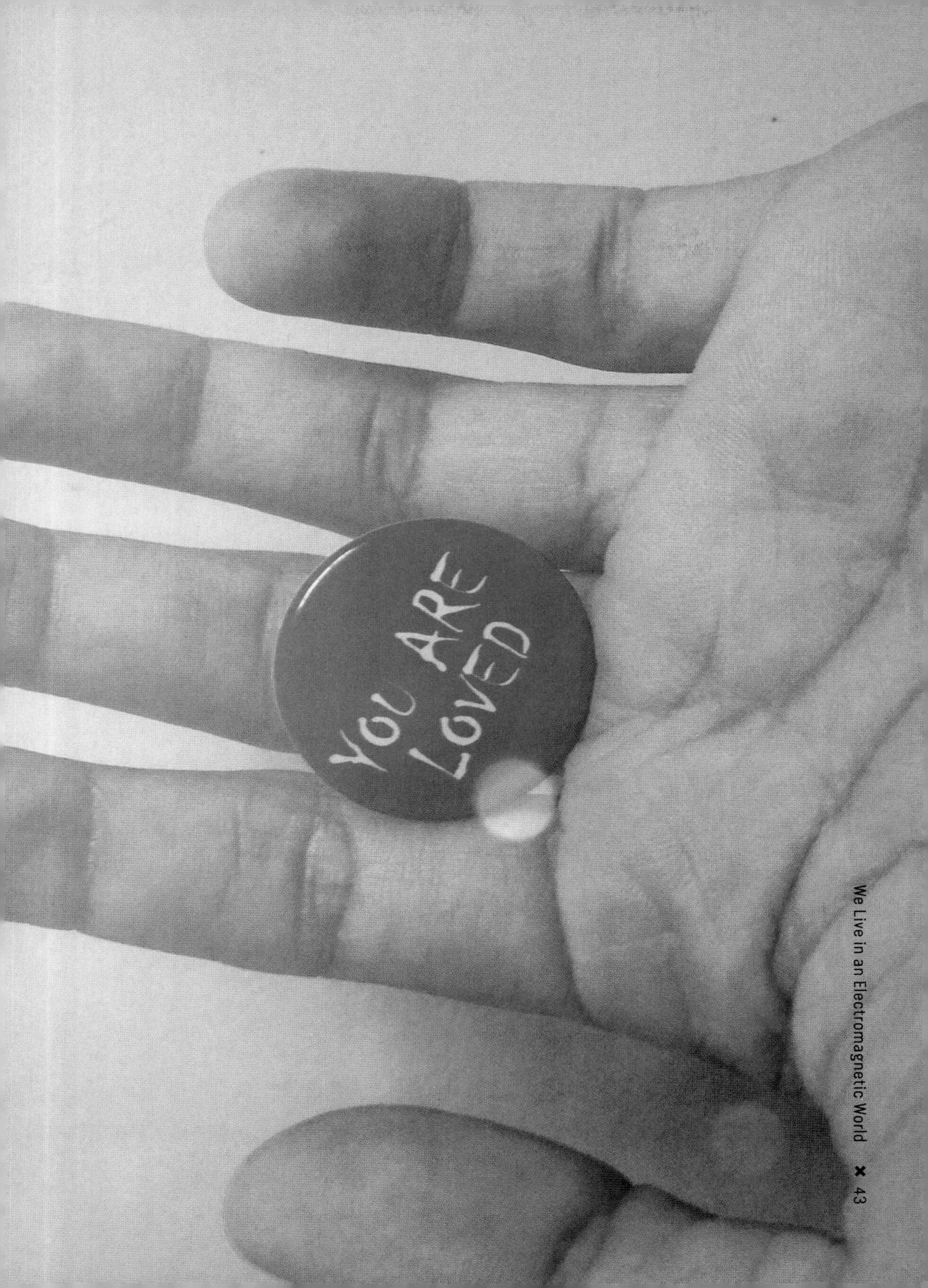
YOU ARE LOVED

Emotions Are Energy

HeartMath has performed several studies that show the magnetic signals generated by the heart can affect individuals around us, from the detection and measurement of cardiac energy exchange between people; to heart-brain synchronization during nonphysical contact; to the biomagnetic communication between people and animals, such as a boy and his dog. These are fascinating findings; we encourage you to investigate further at HeartMath.org.

Emotions are biological, chemical, hormonal, behavioral, and energetic expressions of the state you are in. Emotions are somewhat impermanent; rarely do you experience a long-lasting emotion because of the capacity to move from one emotion to another. The emerging theory among the researchers at HeartMath is that emotions can also be vibrational and that they are shared in the frequency fields we generate.

Neuroscientist and pharmacologist Candace Pert, whom Sputnik Futures interviewed in 2002, was a leading researcher in the "bodymind," the idea that the body and the mind are actually part of a linked system. She is best known for her pivotal role in the discovery of opiate receptors—molecules that unlock cells in the brain so that morphine and other opiates, including the body's natural opiate, endorphins, can enter.

After years of studying the form and function of neuropeptides (tiny bits of protein that consist of strings of amino acids), Pert concluded that they are responsible for a range of our emotions. She looked at how all cells in the body communicate with each other and how cells have receptor sites that receive neuropeptides. The kinds of neuropeptides available to cells are constantly changing, and can influence variations in your emotions throughout the day. This concept that we have molecules of emotion nullified the prevailing idea that the mind has power over the body.[63] According to Pert's research, the body and the mind cannot be separate: "Instead, emotions are the nexus between mind and matter, going back and forth between the two and influencing both."[64]

We Can Catch Each Other's Emotions Like We Catch a Cold

According to social-network expert Nicholas A. Christakis, emotions may be contagious. Our social emotions act like a biological meme, connecting us and going viral through our networks of friends and our friends' friends. Christakis has been at the forefront of proving that our emotions are contagious through our networks, and that we are hardwired to connect through emotions and behavioral influences.[65] The seed of this theory began with the Framingham Heart Study, in which Christakis and his research partner James H. Fowler conducted a longitudinal social-network analysis over twenty years to evaluate whether happiness can spread from person to person and whether niches of happiness form within social networks. They used as their sample more than four thousand individuals within the Framingham Heart Study social network, which followed participants from 1983 to 2003 (the Framingham Heart Study began in 1948, and Christakis and Fowler mined data over a twenty-year span). They reconstructed the social fabric in which individuals were enmeshed (family, close friends, or friends of friends) and analyzed the relationship between these social networks and health, looking at indicators such as the spread of obesity and self-reported moods. The researchers concluded that people's happiness depends on the happiness of others to whom they are connected, and their research ultimately suggested that we should think of happiness, like health, as a collective phenomenon.[66]

We Need to Get a Grip on Our Electrosmog!

So we have discovered the dynamic ways we are tapping into the healing and communication potential of electromagnetic fields, around us and in us. But there is a major disruption quietly invading our electromagnetic fields.

The hyperconnected, electrically charged, wireless world we created in less than a century has resulted in an overabundance of exposure to artificial EMFs. And with the proliferation of the Internet of Things and 5G, everything from public park benches to your kid's stuffed animals are soon set to have an IP address, speaking to each other—and to us—in this invisible smog of electromagnetic frequency. The number of EMFs that humans, animals, and the entire natural world are exposed to is set to increase exponentially.

Our environment is being bombarded more and more with man-made electromagnetic fields. WiFi and 5G are an expected part of an urban landscape, mobile phones rarely slip out of our hands, and our homes are packed with electric appliances of every sort. Electromagnetic (EM) pollution is a new public-health issue, sometimes called electrosensitivity, or ES. There are some interesting arguments—and yes, some denial—that EM pollution is a real health threat. But even Robert O. Becker warned in his second book, *Cross Currents*, that he felt there are dangers of these man-made electrical fields, generated by things such as cell phones and the massive high-tension power lines that dot every landscape inhabited by humans in most corners of the world. Becker writes that industry and government have failed to heed research that shows such fields can be harmful, that even small currents have effects on living tissue.[67]

Electrosmog is a man-made radiation pollutant that includes technology-generating EMFs, electromagnetic radiation (EMR), and microwave radiation that are not natural frequencies. Just like the Earth's EMF, you cannot see or smell these man-made frequencies, but the cells of the body do feel them, and some people suffer more than others with ES or electromagnetic hypersensitivity (EHS). Think about it—over five billion people on the planet today have a mobile phone that, when turned on, emits a deluge of man-made EMFs. Wireless equipment like mobile towers, WiFi, tablets, and laptops emit electrosmog, which we can't escape.

In a September 2019 article in *Down to Earth*, Suresh Karve and Milind Bembalkar cited some alarming facts about electrosmog. One interesting point comes from David Carpenter of the University of Albany School of Public Health, who claimed in the December 1, 2018, edition of *The Lancet* that electromagnetic radiation has increased by one quintillion (one followed by eighteen zeros) times in the past decade and poses a great danger to life on Earth. The culprits of this electromagnetic pollution—mobile phones, antennae of mobile towers, WiFi, and other wireless equipment—work

on frequencies ranging from 700 megahertz (MHz) to 2.8 gigahertz (GHz), and the proposed new arrival of 5G is supposed to work on a frequency of 30 GHz to 300 GHz.[68] That is a considerable increase in electromagnetic frequencies about to be unleashed in our environment.

As we've discussed, Earth has a natural electromagnetic pulse that life forms rely on for navigation and for development of tissues. Research now shows that this pulse has naturally decreased over the previous thousands of years—and that we are accelerating that decline with electrosmog from cell phones, computers, microwaves, etc., which puts undue stress on our cells.[69]

The effect of electromagnetic waves on living creatures has been controversial due to studies with contradictory results. However, in 2011, the International Agency for Research on Cancer (IARC) of the World Health Organization (WHO) designated mobile phone RF-EMFs as Group 2B: possibly carcinogenic to humans.[70] Given that most people, including young children, use mobile phones, the possibility that we might be regularly exposed to a considerable amount of electromagnetic waves greatly increased social interest in the impact of RF-EMF exposure.[71] NGOs such as the Swiss Foundation for Research on Information Technologies in Society (IT'IS) research the safety of emerging electromagnetic technologies such as 5G. One physical effect that this high-power density of 5G can produce is heating the skin (much like a microwave heats food), which could cause thermal damage. According to Esra Neufeld, a scientist and consultant with IT'IS, the current standards for 5G do not prevent thermal damage to the skin. Some countries are already taking precautions. France and Cyprus have prohibited WiFi in kindergartens and restricted its use in primary schools.[72]

According to the WHO, a few individuals per million suffer from EHS, with higher rates in Sweden, Germany, and Denmark. A 2005 WHO fact sheet on electromagnetic hypersensitivity lists the symptoms most commonly experienced, including dermatological symptoms (redness, tingling, and burning sensations) as well as neurasthenic and vegetative symptoms (fatigue, tiredness, concentration difficulties, dizziness, nausea, heart palpitation, and digestive disturbances).[73] In 2016, the European Academy for Environmental Medicine prepared a detailed protocol to be followed for treatment of adverse effects of EMF. It covers prevention, diagnosis, and treatment for such patients.[74]

Still, it's not all doomsday news. NASA, as well as scientists and technologists, have learned to co-opt Earth's natural pulsed electromagnetic field (PEMF) good frequencies to tune us with body-friendly vibrations that affect healing and cellular regeneration. A new wave of spa, medical, and wearable device treatments are matching the intensity of these good frequencies to help us heal and regenerate. (More on that in chapter 4.)

Lo and Behold: Reveries of the Connected World, a documentary by Werner Herzog, features interviews with the residents of Green Bank, West Virginia, a town with no cell phone towers that has drawn a community of people who want to live in an internet-free environment to help fight their "electrosensitivity."

—*Wired*[75]

The Faraday Phenomenon

We owe a lot of the understanding of the effects of electromagnetism to an English scientist named Michael Faraday—you might recognize his name if you are aware of the Faraday cage, and if not, a quick search on Google will turn up more than half a million sites referencing this remarkable invention, which is inspiring a new generation of EMI shielding gear.

Faraday's discoveries include the principles underlying electromagnetic induction, diamagnetism, and electrolysis. In his now famous experiment in London in January 1836, he built a twelve-foot-square "cage" made of a wooden frame, built on four glass supports with paper walls and wire mesh. Faraday supposedly stepped inside and electrified it, living in the cage for two full days. Using electrometers, candles, and a large brass ball on a white silk thread, Faraday explored the nature of electrical charge, transforming how scientists view electricity.

His work is still the basis of new research on how to protect us from electromagnetic pollution, and these days companies are touting products based on Faraday's technology, offering everything from bags, to curtains and wallpaper, to radio wave-proof underpants that claim to stop mobile-phone and WiFi radiation and shield devices from electronic snooping. The aim is to shield and protect you from excessive and unwanted microwave radiation that is now raining on you almost everywhere you go.

In 1996, microwave radiation (MWR) from wireless devices was declared a possible human carcinogen, with children identified as the most at-risk group. According to a review titled "Why Children Absorb More Microwave Radiation Than Adults: The Consequences," published in the December 2014 issue of the *Journal of Microscopy and Ultrastructure*, children absorb more MWR than adults because their brain tissues are more absorbent and their relative size is smaller. The study reported that the absorbed MWR penetrated proportionally deeper into the brains of children ages five to ten compared to adults' brains. [76]

There exists a 20 cm distance

rule for mobile devices, tablets, and laptop computers. For purposes of these requirements, mobile devices are defined by the FCC as "transmitters designed to be used in other than fixed locations and to generally be used in such a way that a separation distance of at least 20 cm is normally maintained between radiating structures and the body of the user or nearby persons."[77]

More and more tablets are used by young children in schools, a trend that seems unlikely to change anytime soon, especially given the new home-school technologies used by educators during the COVID-19 pandemic, requiring extensive use of a mobile computer, laptop, or tablet. This unavoidable use of a computer contradicts the 20 cm distance warning, as children have shorter arms that do not allow them to hold devices 20 cm from their bodies. It is also common these days to see toddlers with a mobile phone or tablet in their lap, often learning to swipe before they learn to speak. According to "Why Children Absorb More Microwave Radiation Than Adults: The Consequences," any MWR-emitting toys should ideally not be used by infants and toddlers. The researchers cite dozens of studies with health claims against MWR exposure, from sperm damage to breast cancer to parotid gland tumors—a read that is both fascinating and depressing and exposes how utterly dependent we have become on our wireless devices.[78]

Silver Nanowire Shield

Researchers at King Abdullah University of Science and Technology in Saudi Arabia have developed a stretchable, electrically conductive mesh of silver nanowires that could be used to protect against electromagnetic radiation. They designed a stretchable material with special electromagnetic absorbers using their transparent ink, which can absorb more than 90 percent of the electromagnetic signals in a specific frequency band. One potential application for the shielding material is on incubators for newborn babies.

—*The Engineer*

EMF-Blocking Fashion

Lambs EMF Blocking Underwear uses WaveStopper technology, a patented silver-lined wire mesh fabric woven directly into cotton fabric blocking 99 percent of harmful microwaves commonly caused by WiFi radiation, Bluetooth, and cellphone radiation. ASTARA women's luxury wellness footwear (launched by a former employee of Jimmy Choo) was designed based on the concept of "Earthing"—reconnecting with Earth's energy for balance and protection. Each shoe is built with a component attuned to the frequency of 7.83, which is the Earth's resonance, and the frequency of the Schumann resonances, which have been shown to positively influence health, performance, and overall well-being in humans.

—*Forbes* [79]

Frequent Travelers Zapped with Radiation

Those who spend a lot of time in-flight face a high exposure to cosmic radiation, which can easily penetrate human skin and poses a risk to our health. The Centers for Disease Control and Prevention (CDC) has classified airline crew members as "radiation workers" based on the dangerously high doses experienced.

—Tech Insider

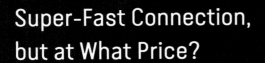

Super-Fast Connection, but at What Price?

The greatest polluting element in the Earth's environment is the proliferation of electromagnetic fields. I consider that to be a far greater threat on a global scale than warming, or the increase of chemical elements in the environment.

—Robert O. Becker (1923–2008), orthopedic surgeon, researcher in electromedicine, author of *The Body Electric* [80]

Global action is under way to both fight and raise awareness of this public health threat. The advisers to the international EMF scientist appeal, representing 248 scientists from forty-two nations, requested that the United Nations Environment Program (UNEP) reassess the potential biological impacts of next-generation 4G and 5G telecommunication technologies on plants, animals, and humans.

The urgency is due to this critical transition time as technology companies are wiring up the world with new antennae that will densely populate residential neighborhoods. These antennae will be using much higher frequencies, with potentially greater biologically disruptive pulsations and more dangerous signaling characteristics that will be transmitting to and from equipment outside and inside of homes as well as schools and other public buildings.

Ronald Melnick, PhD, adviser to the appeal and former scientist at the National Toxicology Program (NTP) within the National Institutes of Health (NIH), managed the design and development of the NTP's recently published $30 million animal study, which showed a clear link between radio frequency radiation (RFR) and cancer. In the report he states that "results from the NTP study show that the previously held assumption that radiofrequency radiation cannot cause cancer or other adverse health effects is clearly wrong." Policymakers the world over should take note.[81]

One of the first studies linking magnetic fields from power lines to adverse effects on human health was published in 1979 by two Denver researchers: the late Nancy Wertheimer, PhD, and physicist Ed Leeper. Based on Wertheimer's field studies of childhood cancers in the Denver-Boulder area, they reported that children who lived one or two houses from what are called step-down transformers (the barrel-shaped devices mounted on the power poles in your neighborhood) had a two- to threefold increase in childhood cancers, specifically leukemia and brain tumors.[82] In 1986, a similar study conducted at the University of North Carolina at Chapel Hill confirmed their findings.[83]

Dr. Wertheimer was perhaps one of the pioneers in understanding the effects of EMF on

From 5G to 6G at THz Frequencies

While the world is still waiting for 5G that introduces mm wave frequencies to give our automated homes, factories, and cars the bandwidths they need to operate without us, a collaboration of tech businesses and research institutes is working on a terahertz (THz) frequency band related to the sixth-generation wireless mobile communication (6G). What will this mean? Wireless transmitting and receiving systems operating between 270 and 320 GHz. THz frequencies for 6G are estimated to become commercial in the next eight to ten years.

—eeNews Europe

WiFi Angels

We all need one—but this is an app to help you sense electromagnetic radiation by turning the WiFi networks that surround you into a choir of singing angels. Every network you are in is translated into a singing angel. Together they create a choir.

—NextNature.net

public health. A member of the Bioelectromagnetics Society and an epidemiologist, Dr. Wertheimer dedicated much of her life's work to researching the exposure to electromagnetic fields involving power lines and any ensuing health effects. Sadly, her research was never supported by major grants or contracts, yet the results of her efforts were published in *Bioelectromagnetics*, the *American Journal of Epidemiology*, *Science*, the *International Journal of Epidemiology*, the *Annals of the New York Academy of Science*, the *Journal of Clinical Epidemiology*, the *Health Physics Society Newsletter*, and the *Journal of the National Cancer Institute*.

Dirty Electricity

Here's a question to consider: Are the light bulbs in your home healthy for you? It turns out that certain light bulbs and dimmer switches can contribute to electropollution with potentially toxic electrical frequencies.[84] These types of "dirty electricity" include electrical power lines, wiring in your home, and all the things you plug into electrical sockets. The excess of this dirty electricity radiates in your home and can cause physical reactions such as headaches or heart arrhythmia.

Electromagnetism is beginning to have a physical influence on the way we construct and manage buildings, including our homes, and architects are waking up to this new challenge. It seems that WiFi moves through architectural interiors like a shimmering curtain. Physicist Jason Cole at Imperial College London published a mathematical model showing how curtains of WiFi shimmer and move through a building, passing from room to room and around walls. Your next poltergeist may be a WiFi curtain!

We are just beginning to understand the EMF health effects that

Dreamhomes is a cooperative community development project in Cyberjaya, Malaysia, that integrates wellness and health care into the construction. They offer a harmonizer that is integrated before construction. This harmonizer acts like a full home protective shield, balancing the distortion of geoelectromagnetic energies in the Earth's field and fault lines as well as man-made underground pipes and cables; and employing grounding and shielding technologies used to offer protection and neutralization of the electromagnetic field (EMF).

—MedicalExpo

the proliferation of technology will create. According to an article in the March 2019 issue of *Washington Monthly*, the world is choking on digital pollution, and we must figure out one day how to manage the waste that the internet is dumping on our world, our lives, and our bodies, much like we are cleaning up the waste produced by the Industrial Revolution.[85] As we continue to connect every object imaginable in the Internet of Things, and the next generation of that with the spatial web creating virtual information overlays on physical objects using augmented and mixed reality, we may be potentially creating a bigger, more invisible toxic load than the harmful effects of toxic chemicals, plastics, and heavy metals in the environment.

"Zap-proof" Yourself

Ann Louise Gittleman's book *Zapped: Why Your Cell Phone Shouldn't Be Your Alarm Clock and 1,268 Ways to Outsmart the Hazards of Electronic Pollution* was one of the first step-by-step guides to counteracting the invisible hazards and pollution from electromagnetic exposure, with simple tips on how to "zap-proof" your home.

Simple things Gittleman suggests you do:

Clear your bedroom of electronics. We spend nearly a third of our life sleeping, and some of our best healing occurs during sleep.[86] This is also the room where you're likely to have a bevy of electronics: a TV, radio, alarm clock or radio, cordless phone, etc. Best to turn them all off—or at least move your bed away from a power outlet. And don't use your smartphone as an alarm clock.

Don't put that laptop (or tablet) in your lap. It radiates harmful EMFs whether it's connected to the AC power adapter or not.

Text whenever possible. If you do need to make a phone call, put it on speaker. The farther away it is from your ear (and head), the farther you are from the phone's antenna, and the lower the signal will be.

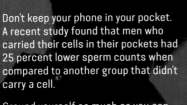

Don't keep your phone in your pocket. A recent study found that men who carried their cells in their pockets had 25 percent lower sperm counts when compared to another group that didn't carry a cell.[87]

Ground yourself as much as you can. Studies show that "Earthing," the practice of getting grounded through bare skin contact with the Earth, helps to promote better sleep and more energy; reduce inflammation and pain; and balance production of cortisol, the stress hormone.[88] The research explains that Earth's negative potential (its electron supply replenishment) helps the human body maintain a stabilizing bioelectrical environment that is important for setting the biological clocks that regulate and create a stable internal bioelectrical environment for the normal functioning of all body systems. The oscillations of Earth's electrons may be important for setting the biological clocks regulating circadian rhythm, and cortisol secretion.[89]

—*Conscious Lifestyle Magazine*

Oska Pulse is a lightweight, wearable PEMF device that is a noninvasive way to help relieve pain. The device emits a diameter field of twenty-two inches and doesn't need to be directly attached to the skin. You can place it near the area of pain and it still will be effective because the PEMF field travels through clothing.

—OskaWellness.com

ActiPatch is also a wearable, drug-free device that provides 720 hours of pain relief therapy through an electrical patch using electromagnetic pulse therapy.

—Source: ActiPatch.com

Not All EMF Is Bad for You

We can thank both the Russian space program and NASA for the technology behind pulsed electromagnetic field (PEMF) devices that deliver low pulsing electromagnetic waves at precise frequencies without the use of heat. Both space programs noticed the physiological and psychological deteriorations of astronauts in space, including bone deterioration, tissue deterioration, and depression, even for those traveling in space for relatively short periods of time. NASA found that the lack of Earth's magnetic field was making astronauts ill and draining them of their energy, so they modified a rotating wall-vessel machine that simulates the weightless environment of space capsules by adding an electromagnetic-field coil to observe the effects of low-frequency, low-intensity, and rapidly varying PEMF signals.[90] Those observations prompted further research that led to the development of PEMF devices that replace the natural PEMF frequencies of Earth with similar signals generated by technology. As an adjunct to the NASA tissue-repair patent, pulsed magnetic fields were also used to stimulate the growth of stem cells.[91]

PEMF therapy was approved by the FDA in 1979 specifically to heal nonunion fractures,[92] and has been used by NASA, the VA, Johns Hopkins, and the Mayo Clinic. There are several portable and at-home therapeutic devices today that help to reduce pain, swelling, and inflammation; to improve circulation, muscle function, and tissue oxygenation; and in some cases to fight depression.[93]

The Electro Basis of the Twenty-One (or More) Human Senses

What constitutes a sense is a still a matter of some debate. As Bruce Durie wrote in his article "Senses Special: Doors of Perception" for *New Scientist*, "When we talk of senses, what we really mean are feelings or perceptions." Perception is like the "added value" our brain gives to the sensory data we take in. Perception may be why people sense differently and may be a combination of different sensory inputs.

Humans have a multitude of senses beyond the basic five. We have a greater ability to detect other stimuli through sensory modalities and receptors that inform one another and give rise to a sense. The modern sensory catalog includes new sense receptors in the muscles, tendons, and joints, which give rise to the kinesthetic sense of motion, as well as the receptors in the digestive tract that help determine experiences such as hunger and thirst.[94]

In fact, by some estimations we have twenty-one or more sensory modalities that help us experience the world. These include temperature (thermoception), the sense of space or kinesthetic sense (proprioception), pain (nociception), and balance (equilibrioception), to name a few. Physiologically, there are electrical signals generated by our sensory cells and afferent nerve fibers that inform our sensory mechanisms.

Animals also have receptors to sense the world around them, varying greatly among species.

Some animals may also take in and interpret sensory stimuli in very different ways. Birds, fish, and other species are able to sense magnetic fields and use it for navigation, a sensory mechanism called magnetoreception. A study published online on March 18, 2019, in *eNeuro* suggests that humans also have a sense of magnetoreception.[95]

Researchers from the United States and Japan recently found that people do subconsciously respond to Earth's magnetic field.[96] The team exposed people to an Earth-strength magnetic field pointed in different directions in the lab, and discovered distinct brain alpha wave patterns in response to rotating the field in a certain way. The team measured alpha waves because of how they react in EEG readings, being dominant when a person is sitting idle, and fading when someone receives sensory input, such as a sound or a touch. The researchers found that when they pointed the magnetic field toward the floor in front of a participant facing north (the direction that Earth's magnetic field points in the Northern Hemisphere) and then swiveled the field counterclockwise from northeast to northwest, it triggered about a 25 percent dip in the amplitude of alpha waves, meaning that the participant subconsciously sensed the field (remember, a dip in the alpha waves happens when we receive a sensory input).

A study led by Joseph Kirschvink, a neurobiologist and geophysicist at Caltech, and his colleagues indicates that we still don't know exactly how our brains detect Earth's magnetic field. One theory is that the brainwave patterns may be influenced by sensory cells containing a magnetic mineral called magnetite, which has been found in magnetoreceptive trout as well as in the human brain. While it is an exciting discovery of our brains sensing magnetic fields, researchers still don't know what our minds might use that information for.[97]

We at Alice have a good idea—let's unplug, go grounding, channel our emotions, do some mindful moving of qigong, recharge with Reiki, and migrate with the animals. Vacation, anyone?

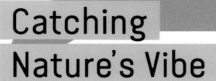

Catching Nature's Vibe

The Wild, Wonderful World of Nature's Internet

Have you heard? Trees just might talk to one another! Research suggests that trees don't just compete for survival, they also cooperate and share resources using underground fungi networks.[1] Hardwired to connect, nature may have created the first electrical grid. In meadows, marshes, and river bottoms, electroactive bacteria are busy building an infrastructure to shuttle subtle frequencies to help control the chemistry of the Earth.

Plants and animals, including humans, feel sound as well as hear it, and some of the most meaningful audio communication happens at frequencies that people can't hear. Frequency is the speed of vibrations, and it can travel in waves where the highs and the lows determine the pitch of sound. So sound is vibrational frequency we can hear, and nature can too. Nature has a unique ability to use frequency and sounds for communicating, and we are learning how to listen to its magical chorus. Elephants use low-frequency rumbles to find family or a mate across long distances. Whales do it too.

Nature Reciprocates

Nature can teach us about cooperation and collaboration. Research suggests that trees share resources, using underground fungi networks. Ecologist Suzanne Simard, at the University of British Columbia, studied how shaded fir trees receive carbon in the summer from sun-laden birch trees, while in the fall firs reciprocate. The birch receives carbon from the fir trees. This exchange takes place through an underground mycorrhizal network—a symbiotic association between a fungus and the roots of its host plant.
—*Yale Environment 360*[2]

Nature's Archives

An ancient kauri tree that contains a record of a reversal of Earth's magnetic field has been discovered in northern New Zealand during excavation work for a geothermal power plant. Scientists surmise that the tree may have lived for seventeen hundred years before the Earth's magnetic field flipped about forty-two thousand years ago.
—*Newsweek*[3]

A rhizome is a subterranean plant stem that sends out roots and shoots from its nodes growing horizontally underground, but also pivots, allowing new shoots to grow upward. The rhizome is often used as a network model of how nature spreads and grows.[4] The philosophers Gilles Deleuze and Félix Guattari, in their project *Capitalism and Schizophrenia* (1972–1980), used the rhizome as a metaphor for an "image of thought," or how cultural thoughts spread.[5]

While we're on the topic of whales, neuroscientists have recently shown that memory recall is coordinated through slow-frequency, thrumming rhythms called theta waves (which are more pronounced during NREM, the dreamless, non-rapid eye movement stage of sleep), and that these slow-frequency rhythms are similar to the pulsing songs shared among humpback whales.[6]

Many birds have what is musically known as "absolute pitch"— the ability to determine exactly what key they sing in without reference to other sounds.[7] For example, if you hear a bird singing a song in G major, you will find it singing the same song in G major again and again. A fun fact: Mozart owned a starling that would sing to him every day, pro- viding Mozart with a melody for his Piano Concerto in, of course, G Major.[8]

Like a computer, nature has a memory that can be stored and accessed. For example, stem cells can memorize the form and function of all cells, allowing blood cells to become muscle cells and so on. In fact, memory may be an inherent property of all matter and space. The late physicist John Archibald Wheeler mentioned to us during an interview in 2002 that he kept on his windowsill a rock from the garden of the Platonic Academy, founded by Plato in about 428 BC in ancient Athens, and that it was his wish that one day there could be a mechanism that could unlock its sounds so that he could hear the discussions between Aristotle and Plato. Perhaps, if we are "in tune" with the Earth, we, as a species, like our kin the rhizomatic tree, can resonate with the collective. In other words, this is our brain on nature.

The Earth Is Alive

My great hope is to understand how the physical world came into being and how it works, and where we fit into it all. I have on the windowsill of a cottage in Maine, a rock which comes from the garden of the Academy in ancient Athens which must have heard the discussions of Plato and Aristotle as they walked back and forth. All I need is some mechanism I can put that rock in which will bring forth a sound.

–John Archibald Wheeler, theoretical physicist, "Father of the Black Hole," Sputnik Futures interview, 2002

Nature's Electric "Wires"

We can hardly write a chapter on nature's internet without starting with efficient, elegant, and electric bacteria. After all, electroactive bacteria were running current through "wires" long before Thomas Edison was born. Electroactive bacteria are nature's power grid and support Earth's chemistry.[9]

In the mid-1980s scientists found that mud is electrical. Well, not the soil, but the microbes that dwell in it. Dr. John Stolz, a microbiologist at Duquesne University in Pittsburgh, and Dr. Derek Lovley, now a microbiologist at the University of Massachusetts Amherst, unearthed a microbe called *Geobacter metallireducens* at the bottom of the Potomac River, where there was little oxygen, which the microbes needed to survive. The *Geobacter* need carbon compounds for survival, breaking down these compounds to transfer them to oxygen, creating water molecules. With little oxygen supply at the riverbed, these clever microbes transferred

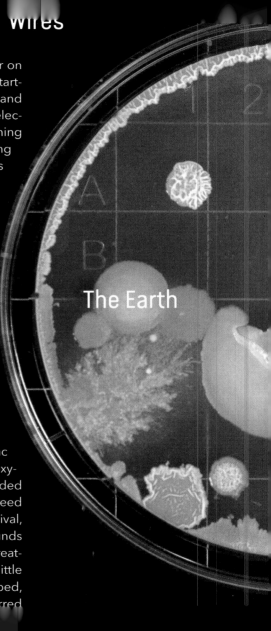

The Earth

Is Electric

their electrons to iron oxide, or rust, which helped rust to then turn into another iron compound, magnetite. Magnetite is the most magnetic of all the naturally occurring minerals on Earth.[10] It is also the compound by which the Greeks first realized the electromagnetic effects of iron ore—how the mineral reacted to electromagnetic fields. As we covered in chapter 1, electromagnetic fields are created by an electrical flow. By helping rust turn into magnetite, electrobacteria are basically generating a current. We also have traces of magnetite in our brain, mainly in the frontal, parietal, occipital, and temporal lobes, which may help tune us to the navigational sense that most animals possess, but more on that later.[11]

Dr. Lovley and his colleagues continued investigating the conductive power of *Geobacter* and found that it was in many different soils, not just riverbeds. In their research the team discovered that *Geobacter* could sense rust in its surroundings, and it would respond by growing hair-like "wires" on it, known as pili. They subsequently found that the pili acted like living wires, sensing the rust around them, plugging directly into the rust, and then extending their "wires" to plug into other species of microbes. In one experiment, Dr. Lovley and his team found that when they used an electrified probe to touch a pilus plucked from the bacteria, the current shot down the hair.[12]

It seems that *Geobacter* weren't the only electrifying kids in town. At about the same time as Dr. Lovley's discovery, a Danish microbiologist, Lars Peter Nielsen, stumbled on a different kind of microbial wire. Also drawing his sample from mud, he observed different chemical reactions by its microbes. These microbes built a vertical "wire" about two inches long and made of a stack of cells encased in a conductive protein sheath, which ran up through the mud. Nielsen and his student Christian Pfeffer found that the bacteria's "wires" aligned themselves into living electrical "cables"[13] similar to twisted bundles of fibers we use in most electronics, surrounded by an insulating sheath.

In 2019, a team of researchers from Washington State University captured bacteria that eat and breathe electricity from four hot pools in the Heart Lake Geyser Basin area of Yellowstone National Park. They developed a new strategy to replicate geothermal features such as hot springs to enrich heat-loving bacteria in their natural environment. The team is one of many other research groups looking at the potential of how special bacteria can "eat" pollution by converting toxic pollutants into less harmful substances and generate electricity in the process.[14] In other words, bacteria may one day be our most natural environmental cleaner and energy producer.

These discoveries are not the first known among electrical bacteria. That title belongs to the industrious cyanobacteria, often called "blue-green algae." Cyanobacteria are some of the oldest and most important microbes on Earth, estimated to be more than 3.5 billion years old. They mainly live in water and, through photosynthesis, manufacture their own food, and help give plants life. It is their vital energetic process of photosynthesis that has currently made cyanobacteria a focus in alternative energy research as some of the most promising feedstocks for bioenergy generation.[15]

Now researchers are finding ways to partner these electrical powerhouses with mushrooms, hoping to understand new ways to generate small amounts of electricity, similar to how biomass (crops such as switchgrass, or waste materials such as wheat straw) is being used today as an organic and alternative fuel source.[16] A team from Stevens Institute of Technology in Hoboken, New Jersey, created a relationship between button mushrooms (*Agaricus bisporus*) and cyanobacteria. To do this, they 3-D printed onto the cap of a living mushroom a pattern like a branch, using electronic ink that contained graphene nanoribbons. The team then printed a bioink containing cyanobacteria and added it onto the same mushroom cap, in a spiral pattern, interweaving it with the electronic ink pattern at multiple points.

What they found was a symbiotic exchange through electrons at the sites where the two printed inks intersected on the mushroom cap, making the mushroom "bionic" and energy-producing. Led by Dr. Manu Mannoor and Sudeep Joshi, a postdoctoral fellow in Mannoor's lab, the team discovered that shining a light on the mushroom activated the cyanobacterial photosynthesis, generating a current of about sixty-five nanoamps. Although that is a nano amount of energy (barely enough to power anything), the researchers put together a network of these bionic mushrooms, showing that the more densely packed together they are, the more electricity they produce—potentially enough to power an LED.[17] The researchers are working on ways to generate even more energy.[18] (More on the magical power of microbes in our future book on bacteria.)

The Internet of Fungi

In nature, it seems that the chattiest species are fungi, showing off through their fruiting bodies (mushrooms) and communicating through their roots (mycelia). These mycelia are the transmitters, connecting not just with other fungi, but also with the roots of other plants, trees, and weeds. As science writer and mycologist Nic Fleming calls it in his November 2014 BBC News article, "it's an information superhighway" right under your feet. Through this "wood wide web," the fungi network with their neighbors near and far, sharing nutrients while acting like homeland security, sending toxins to unwelcome plant intruders.[19]

Roughly 90 percent of vascular land plants are in mutually beneficial relationships with fungi, colonizing the roots of plants in a symbiotic partnership called "mycorrhiza." In these networked associations, plants provide fungi with food in the form of carbohydrates, and the fungi help the plants drink water and consume nutrients.[20]

Plants and fungi also have each other's back, sending warning signals through an energetic exchange in a cross-species social network of collaboration. Fans of the 2009 James Cameron film *Avatar* should recall how all the trees were signaling and responding to one another in the Pandora forest, which the character Dr. Grace Augustine explains as "some kind of electrochemical communication between the roots and the trees."[21]

The mycology expert Paul Stamets, whom Sputnik Futures interviewed in 2002, called the mycorrhizae "Earth's natural internet." Stamets has been studying thousands of specimens of the Pacific Northwest's native fungal genome mycelium, remarking how "waves of mycelial networks intersect and permeate through one another." In "Earth's Natural Internet," an article he published in *Whole Earth* magazine in 1999, Stamets observed that mycelium have a consciousness with which humans might be able to communicate, and that "through cross-species interfacing, we may one day exchange information with these sentient cellular networks."[22]

Earth's Energy Grid

Deep in the philosophy of many ancient cultures is the understanding that Earth has an energy. Certain mountains, rivers, and other natural sites are often referred to as having spiritual power, or a flow of energy that the ancients revered. The ancient Chinese called Earth's energy flows "dragon lines," South Americans call them "spirit lines," and in the West we know them as "ley lines." Ancient cultures were highly aware of those subtle energies, building temples, pyramids, stone monuments, and burial and other sacred sites on or around what they believed to be an energy or spiritual spot.[23] Acting like the energetic veins of the planet, ley lines were believed to radiate subtle energy, but the existence or even the subtle power of ley

lines are controversial in science, as these lines are not directly detected with most modern instruments.[24]

Ley lines refer to straight alignments drawn between various historic structures and prominent landmarks, an idea first proposed in 1921 by English antiquarian Alfred Watkins after visiting a friend in Blackwardine, an English town, where Watkins noticed that a number of the town's churches and historical structures appeared to form a straight line that connected them. He drew a map and was able to connect, across the countryside, the churches, crossroads, and burial stones in straight alignments. He felt at the time that it was some ancient form of tracks or trading route.[25]

Ernst Hartmann, a German medical doctor, began studying ley lines after World War II and discovered a series of energetically charged lines that run north to south and east to west—what is known as the Hartmann grid, often referred to as a "global grid."[26] In his book *Krankheit als Standortproblem* (*Illness as a Location Problem*), Hartmann explains that this energy grid corresponds with Earth's magnetic field. He pro-

poses that there is an order to this grid, in the form of lines that measure eight to twelve inches wide. These lines have a different spacing between them based on the direction they run, sort of like the grid system in New York City: lines that run east to west are approximately ten feet apart, but those that run north to south are about six feet apart. These measures are not fixed widths and vary immensely depending on the geographical location. In his book, Hartmann outlines ways in which the character of a local grid can influence disease.[27] Advocates of energy medicine have described the health effects of the Hartmann Grid using the Chinese terms yin and yang. The north-to-south lines are the yin, and represent a cold energy that corresponds with winter, acts slowly, and can be related to cramps and forms of rheumatism. The east-to-west lines are the yang, which are hot, rapidly acting energy related to fire and linked to inflammations. The intersections of these lines, whether positive or negative, are sensitive to the rhythms of the hours and the seasons.[28]

But the history and lore of some ancient cultures around

the world hold a belief in a deeper purpose behind the lines. In his article "The Ley of the Land," David Newnham dives into the folklore and modern theories of ley lines, explaining how Watkins's findings became the seminal work on ley lines, and discussing the societies in England that were dedicated to "ley hunting"—finding the original tracks or points of connection between revered public structures that were active until World War II.[29] There are many books exploring the possibility that since ancient cultures used these tracking lines in spiritual ceremonies, such as burying the dead, and as a direct route between spiritual structures and monuments, that there was, well, a spiritual reasoning for it. In his book *Ley Lines: The Greatest Landscape Mystery*, author Danny Sullivan notes how easy it can be to find ley lines anywhere, yet maintains that some may have been intentional for ancient burials. He explores why many ancient cultures have a common theme of lining up places, whether it be between spiritual structures and ceremonial paths, or astrological alignments with the stars.

The union of spirit and energy and Earth is woven in the ancient texts of many cultures, and theories from the early nineteenth century referenced an omnipresent energy of Earth and ether (air), the "universal life force" (are you with me, baby Yodas?). One noted concept that has been referenced in subculture novels from the beat generation is "orgone," proposed by Wilhelm Reich in the 1930s, and developed after his death by Reich's student Charles Kelley in the 1960s. The origins of the theory are explained as an ordered principle of the universe, a creative substratum in all of nature that builds on earlier hypotheses such as Franz Mesmer's animal-magnetism theory that every living thing had an invisible natural force (1779); Baron Carl von Reichenbach's 1845 theory of a vital energy or life force he called the "Odic force"; and the French philosopher Henri Bergson's view of élan vital (vital force), which he proposed in 1907 as a possible explanation for the evolution and development of organisms.[30]

Although modern technology still has not proven that ley lines exist, we cannot ignore the beliefs and practices of ancient cultures who shared a common premise that there is an energy force—astrological, spiritual, and Earthly—that connects humans, animals, and the Earth. We just may have to look at the way nature communicates to understand this possibility.

May the Force Be with You

The Force is the underlying "energy field created by life that binds the galaxy together" in the popular Star Wars film series. Harnessing the power of the Force gives the Jedi (the light side) and the Sith (the dark side) extraordinary abilities and can direct their actions, such as when the all-knowing Yoda appeals to Luke Skywalker to "feel the Force" around him. The movies overtly play on the tension between the light and the dark sides of the Force, and how we can tether between their lull, as in the once-heroic Darth Vader, who was seduced to become a Sith lord and works to eradicate the empire. Full disclosure here: we at Alice are Star Wars fanatics. And yes, someone on our team has a lightsaber. We can go on and on with the subliminal references of universal energy of light and dark in every Star Wars film, but for now, may the Force guide you Down the Rabbit Hole (chapter 8) for further reading . . .

Nature's "Voiceprint"

Did you know that volcanoes "sing"? That is, they have an infrasonic sound that scientists have been listening to for nearly two decades. Listening to these low "voiceprints" in active volcanoes such as Cotopaxi in Ecuador or Hawaii's Kīlauea helps scientists study what's happening at the surface of an active volcano as well as volcanic eruptions and the movement of volcano-induced mudflows, known as lahars. By mapping a volcano's unique voiceprint, researchers can track any significant changes in activity deep within the volcano.[31]

These sound waves have frequencies far too low for humans to hear, between about 0.01 Hz and 20 Hz. However, it is a frequency audible to some animals, including elephants, who can hear sounds at 14–16 Hz, and some whales can hear infrasonic sounds as low as 7 Hz (in water).[32] The differences in our range of hearing may help explain why certain animals are known to flee for safety before a natural disaster such as a volcanic eruption or a tsunami.

Another explanation may lie in animals' sensitivities to changes in the electromagnetic field. According to the research of quantum geophysicist Motoji Ikeya, certain animals, even aquatic ones such as catfish, react to changes in electrical currents. His research has included regularly monitoring catfish, the most sensitive of the creatures he has tested, in sensing coming disaster.[33]

Researchers have found that plant-borne vibrations provide a wealth of information about the activities of insects on plants, including many ecological and social interactions that depend on the production of the sound waves generated by plants.

—Peggy S. M. Hill, *Vibrational Communication in Animals*[34]

Some two hundred thousand species of insects communicate using substrate vibrations—frequencies generated close to the ground or surface—to locate mates, attract collaborators, or exploit a plant's resources.

—*BioScience*[35]

Bioacoustic Conservation

Recording the sounds of a forest's landscape could unlock secrets about biodiversity and aid conservation efforts around the world, according to a perceptive paper published in *ScienceDaily*. Bioacoustic devices can be programmed to record at specific times or continuously. Bioacoustics can help to effectively enforce policy efforts as well. Many companies are engaged in zero-deforestation efforts, which means that legally they cannot clear large forests. Bioacoustics can quickly and cheaply determine how much forest has been left standing.

—*ScienceDaily* [16]

Nature's Ears

We've learned so far that plants can influence one another through the wood wide web, a network of fungi that connects their roots. Now scientists are researching how plants respond to vibrations, especially when it is the buzzing frequency of insects. Some of these insects can be invaders or "chewers," and plants identify their frequency, creating defensive chemicals. Alternately, when it comes to friendly pollinating insects, plants can hear them using their flowers, and release pollen to help in their tasks.

This vocal dance between plants and insects is the focus of experiments by Lilach Hadany and Yossi Yovel at Tel Aviv University. They showed that some plants can hear the sounds—airborne vibrations—of pollinators and react by rapidly sweetening their nectar. Hadany explained to the *Atlantic* in 2005 that the premise of their theory stems from the fact that animals both make and hear noises, and since plants and animals interact, it seemed "weird to think that plants wouldn't make use of sounds—airborne vibrations" to sense.[37]

If humans and most mammals have ears to hear, what then are a plant's "ears"? Hadany's team found that the ears are actually the flowers themselves. Using lasers, they were able to show that an evening primrose's petals vibrate when hit by the sounds of a bee's wingbeats, producing nectar. When they covered the petals with glass jars, essentially blocking the vibrations from the bees, the nectar never sweetened. They surmised that the flower could act similarly to how our outer ears work, channeling sound farther into the plant.

Some flowers react when bees rapidly produce strong vibrations, causing the inside antlers of certain flowers to release pollen, a process known as buzz pollination or sonication. But the flowers in this study reacted quickly, increasing the concentration of sugar in their nectar by close to 20 percent in just three minutes after hearing the bee sounds. Interestingly, the researchers observed the same result using artificial sounds of a similar frequency, suggesting that, just as important, there was no change in sugar levels in the flowers when exposed to higher-frequency sounds, or no sound at all.[38]

Plants also use vibrations to sense potential invaders. While sound waves have been shown to have an effect on things such as germination and growth in plants,[39] a team of researchers from the University of Missouri have shown that plants can also sense the vibrations from insects that are eating them, defending themselves in response. Led by Rex Cocroft and Heidi Appel, the team recorded the sounds of caterpillars munching on the leaves of a small flowering plant related to cabbage and mustard, supposedly a favorite of caterpillars. The team then played back the sounds of caterpillar chomps for the plants, who sent out extra mustard oils into their leaves—a potential toxin to caterpillars. The researchers did not notice a response from the plant with other sounds, such as wind or insect song, which suggests that plants just may be able to distinguish between bad eating vibrations and regular environmental vibrations that are not harmful.[40]

There is one theory on plants' ability to sense that was documented in experiments by plant biologist Cleve Backster. His theory emerged in the 1970s when he was working as an expert in lie detection with the CIA. On a whim, he hooked up the machine's galvanometer—the sensor that measures small electrical currents—to a leaf of a dracaena, a common houseplant. In humans, the galvanometer measures changes in heart rate or pulse—both potential indicators of stress—during the interrogation process. Backster tested the plant by imagining the dracaena was being set on fire. When he did, the plant reacted with a surge of electrical activity that was registered on the polygraph machine. This surprising discovery prompted Backster to see if there might be other reactions in the plant. For example, he attempted to elicit fear by setting one of the plant's leaves on fire, but before he could even light the match, the polygraph registered an intense reaction from the leaf. When he returned with a matchbox, he saw another reaction appear. The plant seemingly felt that he was about to burn it; therefore it exhibited signs of

fear. To Backster, it was indisputable proof that the plant not only demonstrated fear but also could somehow read his mind or comprehend his intentions. He went as far to say that plants could be conscious.[41]

This plant ESP phenomenon was named the "Backster Effect." Although the scientific community was less moved by his work, Backster continued his experiments until the end of his life. His expanding theory of nonhuman consciousness and studies documented what he called the fundamental interconnection between all living things. The recordings of his experiments can be found in his book *Primary Perception: Biocommunication with Plants, Living Foods, and Human Cells.*[42] His theory of plants having consciousness and emotions is embraced by some naturopaths who work with plant-based therapies. We have been fans of Backster's work for years, and we are hopeful that one day, with the use of more advanced instruments such as magnetoreceptors, we may witness the energy fields of plants.

Cleve Backster explains in his book *The Secret Life of Plants,* originally he had been using [lie detectors] on plants and now he's using it on detached parts of the cells of the body. And he has the same results as he had in plants, namely when the subjects undergo an emotional experience, his or her cells react as well. It doesn't matter how far away they are.

—Ervin László, systems theorist, Sputnik Futures interview, 2002

With more advanced instruments, one thing we are learning is that Mother Nature has a voice, and plants have a melodious language. Plants are musical, and we can hear them play through biosonification devices. One device, the Bamboo, from Music of the Plants, deciphers and registers the impulses and interactions of plants through a device that uses a MIDI (musical instrument digital interface), which transforms the plant's inner movement from a leaf to the root system into music.[43]

PlantWave is another digital

interface that turns a plant's bio-rhythms into music. Developed by multimedia healing artists Joe Patitucci and Alex Tyson of Data Garden, PlantWave acts like an external drive that records the slight electrical variations in a plant via two electrodes placed on the leaves. These variations are graphed as a wave, which is translated into pitch messages that play musical instruments in the form of software that the Data Garden team built for the plant's frequencies to play! What you then have is a playlist of your home or garden plants' music. The PlantWave digital recorder connects wirelessly to your smartphone or any iOS or android device and pairs with the PlantWave app (which you have to

download to access the record-ings) to play hours of live plant music. The result is "a continu-ous stream of pleasing music that gives you a sonic window into the secret life of plants."[44]

The people [Aborigines] themselves say, "Don't you know anything? The plants talk to us." And you begin to think, *Well, maybe it's true.*
—Wade Davis, explorer in residence, *National Geographic*, Sputnik Futures interview, 2002

Well, Hello!

So far in this chapter, we have explored the wild, wonderful communication abilities of the Earth, plants, and the wood wide web of tree roots and fungi, mostly through instruments and studies that we humans use to detect these abilities. But what exactly are these wonderous creatures and plants talking about? Can we get inside the sounds these creatures project to see and hear the world differently? What can we learn from animals that are endowed with different sensory apparatuses?

It may surprise you that birds aren't the only singers on the planet. Whales are known to have a range of sounds that help them communicate and socialize. They employ a series of clicks, whistles, and pulsed calls to assist them in navigating, communicating with each other, and identifying predators. They travel in groups known as "pods," and there are different vocal "dialects" between pods. Many researchers have captured the symphony of their conversations, some apps on Google Play use whale songs to help you sleep, and several YouTube meditation videos use the underwater sound track as part of relaxation music. For your listening pleasure, Spotify has several playlists of whale songs; check out "Whale Sound," featuring ten songs for relaxation.

One of the early pioneers of whale listening is Katy Payne, a

zoologist and bioacoustics biologist who started recording the songs of whales in 1968 with her husband. They started off in Bermuda, using underwater microphones to capture the sounds of humpback whales. After thirty-one years of analyzing the recordings, Payne discovered predictable ways in which the whales changed their songs each season and, with her colleague Linda Guinee, that whales use rhymes in their songs.[45]

The songs of male blue and fin whales, the largest-known animals, can travel more than six hundred miles underwater, making it possible for the whales to communicate many miles across the ocean. But scientists around the world who have been tracking the pitch of whale songs for decades have noticed that the whales have been singing a little flatter than usual. A study in the American Geophysical Union's *Journal of Geophysical Research: Oceans* analyzed more than 1 million songs from three species of large baleen whales: fin, Antarctic blue, and three populations of pygmy blue whales, who are acoustically distinct from other whales. The researchers used six stationary underwater microphones that recorded the calls over six years, from 2010 to 2015, in the southern Indian Ocean, an area spanning roughly 3.5 million square miles.

According to the study, blue whale songs are in the range of the lowest, longest pipes of large cathedral organs. The researchers measured the pitch of selected elements of each species' songs, which had fallen to about 25.6 Hz for the Antarctic blue and 96 Hz for the fin whale by the end of 2015. The study attributes a falling pitch in fin whales and Madagascar pygmy blue whales to either an adjustment to the growing numbers of whales, or changes in the ocean resulting from climate change. The researchers also found that calls from blue whales in the southern Indian Ocean increase in pitch during the summer, probably to be heard over breaking sea ice, according to the study's authors.[46]

Declines in Arctic sea ice cover are influencing the presence of popular marine mammals, impacting not just the migration of these mammals but also affecting the livelihood of local indigenous peoples. Scientists have engaged acoustic monitoring of Arctic marine mammals to understand the pressures of climate change and provide insights to help these animals adjust. A four-year acoustic study by the WCS (Wildlife Conservation Society), Columbia University, Southall Environmental Associates, and the University of Washington listened to how seasons, sea surface temperature, and sea ice influence the movements of bowhead and beluga whales, walrus, and bearded and ribbon seals in the Arctic. By listening to the changes in their sounds, scientists hope to understand any variables in their seasonal presence in light of the challenging Arctic changes, and how ocean noise, from both natural and human sources such as shipping, may affect the behavior and well-being of marine mammals.[47]

Certain Aborigines of Australia have invented a way of communicating with whales that consists of wiping the sweat of the armpit with its distinct aroma and its chemicals, then speaking into one's hand, placing it in the ocean, and sending this chemical/olfactory/auditory message to the whales. It's meaningful for them and they have understood that communication among whales is not visual but is by these chemical and auditory signals. Now, as far as the human language and whether human language can actually be mixed with sweat and odor and transmitted in that way? Well, why can't it be? Why shouldn't it be?
—Constance Classen and David Howes, anthropologists, coauthors of *Ways of Sensing*, Sputnik Futures interview, 2002

How Does Nature Differ from a Magic Show?

There are so many beautiful songs coming from our natural friends that we just can't hear. They swirl and chirrup and pitch all around us like an invisible ultrasonic magic show. Mice sing at a high register—so high that we can't hear it. Bats are echolocators, using sounds they produce to detect objects and for navigation. Their biosonar music is produced as a loud ultrasonic series of slaps and clicks, at about the same volume as a fire alarm, but because they emit between 45,000 Hz and 70,000 Hz, we humans can't hear it. (We can only hear between a measly 20 Hz and 20,000 Hz.) These signals are usually generated from the bat's head, but a new study found that at least three species of bats generate the echolocation signals through the beating of their wings.[48]

Moths are another species of echolocator, and the clicking noise moths make to help them navigate is being used by blind people to help develop a biosonar sense. According to a study published in *PLoS Computational Biology*, some blind people are known to have developed extraordinary skills in echolocation by using mouth clicks, the sounds generated by moths and bats. The researchers looked at the sonic pattern of the clicking sounds and found that there is a spatial path that the sound waves travel once they leave the person's mouth, and that people, though legally blind, could identify shapes surrounding them. One of the study's subjects, Daniel Kish, president of World Access for the Blind, has been using clicking sounds since childhood to interpret the sound that reflected off any surrounding objects into information of the area he was in. To analyze how he and two other subjects were able to navigate through their clicking, the collaborating team of researchers, led by Durham University psychologist Lore Thaler, studied their sounds in an anechoic chamber— a room that totally blocks sound, isolating one in silence. All three of the expert echolocators produced similar bright and brief sounds, high-pitched frequencies at about 2 to 4 kHz, and their acoustics of transmission were consistent in duration (about 3 milliseconds) and directionality, which exceeded the average directionality of speech.[49]

The data from the study is a basis to develop synthetic models of human echolocation, which will help understand the link between the physical properties of acoustic clicking (from sound generation to perception) and human reaction in tasks such as recognizing an object and navigating around it. By mimicking the biosonar communications of our nature, we may one day find a way for the blind to "see" with their ears.

Can You Hear What I Hear?

Every human hears sound differently, and the same can be said for animals, whose varying sounds may serve different purposes. It's all in the pitch of these sounds. There is the high-pitched ultrasound of echolocation that helps bats navigate. There is a midfrequency range that most animals use for communicating in relatively close environments, and some sounds that humans can hear. But there is a magical range below our hearing capacity that is the low infrasound frequencies that animals such as elephants use to communicate over long distances. Infrasound frequencies are sound waves with a frequency below 20 Hz, the lower limit of human audibility. At this decibel, the infrasound frequencies travel long distances without getting reflected off surfaces or objects.[50]

Infrasound is an emerging new field of study in animal behavior. One group, at Cornell Lab's Elephant Listening Project (I can feel you smiling now), has shown that African elephants can communicate over very long distances using infrasound, but has not yet identified what these large, friendly mammals are actually saying. Elephant sounds—especially mating calls—are like rumbles that have many harmonics or multiples of the same frequency, and humans can hear these calls if they are very loud. (If you want to listen, grab your earbuds and visit elephantlisteningproject.org—you're in for a treat!) What's more interesting still is that at higher intensities it is also possible to *feel* infrasound vibrations in various parts of the body.

This is how Katy Payne, the zoologist who first studied the songs of humpback whales, discovered elephant infrasonic communications in 1984 while observing Asian elephants at the Oregon Zoo in Portland, Oregon. What tipped her off was a thrumming vibration in the air that she felt while watching the elephants interact, and realized that she was feeling their conversation. Working further at the zoo, Payne and her colleagues revealed that elephants were indeed making infrasonic calls.[51] Payne and Cornell University acoustic biologists Carl Hopkins and Robert Capranica placed microphones about 5 m away from the elephants, recording them for more than four months, and Payne published the results in her 1999 book *Silent Thunder.* Their work was later confirmed with playback experiments on wild African elephants, using autonomous sound recorders. It was concluded that elephants use their powerful, deep calls in long-distance communication to coordinate group movements and to find mates. In experiments on the savannas of East Africa, they played back recorded elephant calls and observed how the elephants responded to each other's vocalizations over great distances. Payne and her colleague William Langbauer Jr. found that an elephant rumble could reach family members anywhere in a 50 sq. km area from the caller.[52]

Here's another thing about elephants: they can listen to other infrasound in the wild and use it for navigation. Research on elephants in the extremely dry region of Namibia revealed that they may actually listen to the infrasound produced by dis-

tant thunderstorms to find water during drought. Elephants are known to have sudden, seasonal migration movements for reasons largely unknown to researchers. Michael Garstang, an environmental scientist at the University of Virginia, and colleagues conducted a study on nine elephants fitted with GPS sensors and tracked for more than seven years in the dry savanna of northwestern Namibia. The data showed that there were statistically significant changes in movements prior to or near the time of wet episodes within the dry season. While the environmental trigger that caused the elephant excursions remains uncertain, the researchers did find that the rain system generated infrasound, which can travel such distances and be detected by elephants and may possibly be the trigger for such changes in movement.[53]

For years it was unknown how elephants could find others in the vast expanses of the savanna, until it was found that they were emitting very loud (85–95 decibels—about as loud as a motorbike) rumbles as low as 14 Hz.

—*Prospect Magazine*[54]

Further studies have shown that elephants are vigilant responders to alarm-call vocalizations, and send each other specific calls to inform that a calf is arriving, humans are nearby, and even to alert others to angry bees. The organization Save the Elephants and Disney's Animal Kingdom teamed up with scientists at the University of Oxford to study elephant families in two national reserves in Kenya. One of their interesting observations was that the elephants not only made a unique rumbling call warning about bees but also fled from the buzzing sounds by shaking their heads.[55] (Sound familiar? Okay, we may not make a rumbling sound, but you have to admit we may have the same reaction to bees.)

Another unique way elephants communicate over distances is through the ground, as they can separate seismic information from regular background noise. But how can they distinguish between these calls? Research has established that elephant herds can distinguish the contact calls of other herds as being part of their bond group or an outside group.[56] If they can detect who is friend and who is foe, it seems plausible to

Elephants Pay Homage

In 2012 there were several media reports of two herds of wild South African elephants that traveled more than twelve miles to keep a "vigil" at the home of Lawrence Anthony, fondly known as "the Elephant Whisperer," when he passed away. The elephants were rescued and rehabilitated by Lawrence, and his family reported that they had not visited his home for a year and a half prior to his passing. The elephants reportedly stayed for two days before they ventured back into the Zululand bush.

If there ever were a time when we can truly sense the wondrous interconnectedness of all beings, it is when we reflect on the elephants of Thula Thula. A man's heart stops, and hundreds of elephants' hearts are grieving. This man's oh-so-abundantly loving heart offered healing to these elephants, and now, they came to pay loving homage to their friend.
   —Rabbi Leila Gal Berner, PhD, commenting on elephants paying loving homage to their friend; WSFM.com (Australia)[57]

Test how well you can hear infrasound frequencies: The Cornell Lab website shares some fun facts and deep infrasound hearing experiences on their Elephant Whispering Project. Try the quick frequency hearing test with ranges from 30 Hz to 20 Hz to 10 Hz to check out your hearing capacity. And you can listen to the mating calls of forest elephants in the Central African Republic.

—elephantlisteningproject.org[58]

The Earth Rumbles

Earthquakes are another of nature's infrasonic signals that ripple through the ground, sometimes creating a booming sound, and making the crust buckle. A new study finds that Earth's surface acts like a speaker for low-frequency vibrations and can actually transmit that ground frequency into the air, which scientists are hopeful will help with better prediction of earthquakes.

—Live Science[59]

researchers that elephants might be capable of doing the same via the seismic waves transferred through the ground environment. The anatomy of an elephant plays a part in this, from the fatty cushion of its foot to the size of its ears. But what this demonstrates is that a vibration sense is an important tool for survival in many mammal species. We will explore this vibration sense in humans and its detection and healing properties more in chapter 4.

Infrasound is part of seismic signaling to detect prey, avoid or threaten predators, navigate within the environment, and communicate. It may just be that frequency—in this vibration sense and the distance-traveling infrasound—helps with survival, nature helping nature.

Natural infrasound is generated by events such as earthquakes, pounding ocean waves, volcanoes, and severe storms. Other animals besides elephants have shown their ability to detect coming storms from infrasound, and some migratory birds were found to detect coming storms from as much as hundreds of miles away. Researchers led by ecologist Henry Streby at the University of California, Berkeley, in 2014 were tracking a population of golden-winged warblers in the mountains of eastern Tennessee when they discovered that the birds suddenly left their breeding ground and flew south, to the Gulf of Mexico. This spontaneous migration happened two days prior to the arrival of a powerful supercell thunderstorm, the kind of storm that has a persistent rotating updraft that leads to tornadoes. The researchers reported that when the birds made their exit, the storm was still 250 to 560 miles away. They couldn't find any other environmental cues at the time—such as change in atmospheric pressure, temperature, or wind speed, because the storm was still a great distance away—and as a result, they surmised that the birds detected the infrasound of the impending tornadoes.[60]

On December 26, 2004, a seemingly normal, sunny day during the holiday season, an undersea megathrust earthquake that registered a magnitude of 9.1–9.3 megawatts rumbled under the Indian Ocean, unleashing a fast-moving tsunami onto islands of Thailand, Sri Lanka, and Indonesia. It was the world's deadliest recorded tsunami and became known as the Boxing Day Tsunami, taking 230,000 human lives in its wake.[61] The people in these areas had no warning signs, but it seems that the animals did. According to a National Geographic article written just ten days after the disaster, relatively few animals were reported dead.[62]

The article lists several observations of abnormal animal behavior of both wild and domesticated animals, some just an hour before the tsunami hit, speculating that they sensed it was coming. Elephants in Sri Lanka screamed and ran for higher ground, and flamingos in a wildlife sanctuary in Point Calimere, India, fled their low-lying breeding areas. The Indo-Asian News service reported that along the Cuddalore coast of India, where thousands of people perished, buffalo, goats, and dogs were found unharmed. These accounts support the theory that animals can hear or feel the changes in Earth's vibration, and sense approaching disaster before humans ever realize what is about to happen.[63]

The Call of the Wild

We're still learning about the magical playlist of animals, the infrasound "songs" they create to attract mates or warn their friends of prey. Some of these songs are whole-body performances, such as crocodilians performing a "water dance," creating infrasound by making the shallow water "dance" over their backs. When males perform, they usually include head slaps and grunts to attract females, as well as to warn each other of rivals. Tigers, one of the more solitary big cats, also create booming low-frequency sounds to protect their territory from rivals and to attract mates; conservationists think that this may explain how tigers maintain their large hunting territories.[64]

We at Alice in Futureland are animal-loving groupies, and we can go on and on about the songful conversation animals are having that we are just not hearing . . . yet.

"If you are lucky, you might meet an animal that wants to talk to you" claims author Eva Meijer in her book *Animal Languages: Revealing the Secret Conversations of the Natural World*. A Dutch philosopher and artist, Meijer shares examples of how animals do more than just give warnings—like us, they "gossip," share advice, greet each other, and claim their love through "complex languages with grammatical and structural rules." We think of language with its subtext and patterns as humancentric, but it's apparently not. If you are a pet owner, you already suspect that your pet communicates with you through its eye contacts, body movements for a pat of affection, and by the different vocalizations your pet emits that are based on behaviors and expected routines, such as, "Ready for my walk, are you?!" It's a relationship communicated through rewards and interactions day after day.

If We Could Talk to the Animals (and Insects)

Meijer uses examples from the wild of what she implies is a complex and recursive pattern that could be viewed as a form of grammar in the songs of birds and the calls of whales, pointing out the variations of these sounds from animals in different geographic locations as well as different breeds. One example is the scent signals of bees as possibly being a form of grammar based on the hive.[65] Honeybees are interesting creatures, as their communication is based on three sensual systems: touch, scent, and movement. When they greet each other, bees touch antennae to identify each other and to smell the pheromones that identify each unique family hive. The waggling honeybee dance, though, is the most intricate of conversations, and its energetic dance steps describe to the hive members the location of good foraging sites. The hive is believed to "vote" on where to forage by other bee members joining the dance of the communicating bee, an observation that led to the pop-culture term "hive mind." In his book *Honeybee Democracy*, animal behaviorist Thomas D. Seeley shares years of pioneering research on the waggle dance and how each step translates into a specific meaning, concluding that honeybees make decisions collectively and democratically—and have a lot to teach us.

The buzz noise of bees is also a way they talk to each other and to us. Beekeepers know to listen to the intensity of the buzz, as it reflects their mood and if they are stressed or feeling threatened. The quieter the buzz, the calmer the hive is. And that intense high-pitched volume is when they are uneasy.[66]

To some people, the buzz of bees is annoying, but probably

not as annoying as the incessant buzz of mosquitoes. But the irritating acoustics of mosquitoes are due to the aerodynamics of their wings beating to stay airborne while generating and directing the sound to their mates. In other words, they fly and flirt at the same time. The males swiftly flap and truncate their wings far faster and shorter than other winged insects their size, giving them that monotone buzz that gets, well, under your skin before they land on you for a feast. But the real reason for the skin-crawling tone is to connect with females.[67]

Researchers are getting inside this mating buzz to figure out how we can outsmart these pests in a nontoxic way by interrupting their mating chorus to confuse their breeding, and hopefully decrease their population growth in the fight against the West Nile virus; diseases such as malaria and forms of encephalitis; and, of course, to avoid those itchy red bug bites. The goal is to use frequency as a disrupter by playing a sound that interrupts the mosquitoes' sonic mating ritual.[68] To all you bioengineers reading this: Can we think of a wearable buzz diffuser? The world will thank you!

Mouse Speak

DeepSqueak is a new type of AI that decodes what mice and rats are saying by translating the vocals into sonograms, helping researchers visualize and match the different vocal pitches with behaviors and potentially moods. They are interested in discovering the psychiatric health of mice—in this case, laboratory rodents—to detect anxiety and depression. Understanding an animal's emotional wellness through vocalization is a way by which animals can tell us how they are feeling, and ultimately, the social lives of the thirteen hundred rodent-like species.
—*Smithsonian Magazine*[69]

The Animal-Computer Interaction Lab at the Open University (UK) is the world's first research program dedicated to developing "user-centered" technology for dogs and other animals. Working with the UK charity Dogs for the Disabled and designers, the team of animal behavior specialists developed prototype buttons that enable dogs to open doors, switch lights on or off, put the wash on, or raise an alarm if a disabled owner falls ill.
—*The Guardian*[70]

Maybe Dr. Doolittle
Should Be a Woman!

We salute the outstanding women who have pioneered new discoveries in animal communication:

The least I can do is speak out for those who cannot speak for themselves.
–Jane Goodall, primatologist and anthropologist, who at age twenty-six ventured into the wilds of Tanzania to live with, study, and listen to chimpanzees[71]

By golly, they're singing, of all things. They're doing something that we recognize as singing. And so what that has done for me is to make me feel that what lies ahead to be discovered is absolutely limitless. We are not at the pinnacle of human knowledge. We are just beginning.

–Katy Payne, acoustic biologist, who decoded the language of elephants and was among the first scientists to discover that whales are composers of song[72]

Animals simply show their truths energetically, and that incoming message on my side (in a more quantum form) gets drawn up from the unconscious/intuitive reception to my conscious mind—at which point my brain will attach a mental image, word, sensation or emotion to it. That's already the first layer of "interpretation," and some things get lost in translation.

–Anna Breytenbach, professional animal communicator trained through the Assisi International Animal Institute in California who has been practicing for more than a decade in South Africa, Europe, and the United States[73]

Nature Has a Compass

Beyond detecting the presence of friendlies or foes, and the call for mating, nature also has a synchronized symphony of frequency that the tiniest of species (zooplankton and fireworms) to massive humpback whales leverage to chart their migration patterns. They are using other frequencies of Earth, even the universe, tuning into the length of day and night, the phases of the moon, and seasonal changes. For example, bees and ants seem to memorize where they are going based on how the sun tracks across the sky, making "beelines" straight back to cut their travel distance short.

Many animals are thought to have an internal compass—from sea turtles and shrimp, to spiders and crickets, and the most studied of all, the humpback whale. These aquatic beasts follow a trajectory across great distances, are often challenged with extreme currents and storms, yet make their way, rarely deviating off course. And scientists speculate that it all

comes back to geomagnetism—their ability to somehow read Earth's magnetic field. But what biological receptor, exact organ, or antennae—whether external or something dispersed throughout the body—is making this possible is still a mystery. One theory is that there are minute slivers of mag- netite, iron crystals, infused at the cellular level, giving animals the orientation of north and south.[74] This brings us back to the electro- magnetic fields of Earth and bio- magnetism, which we discussed earlier. It is like a silent conductor of nature's rhythms and ebbs and flows.

The Mathematical Beauty of Nature: The Golden Ratio

There is a golden ratio—a similar sequence of proportions—in nature. Fans of *The Da Vinci Code* might recall how author Dan Brown talked about the "divine proportion." In 1498, Leonardo da Vinci created illustrations for a book, *De Divina Proportione* (*On the Divine Proportion*), which revealed a "mystical" mathematical proportion. It's a beautiful hidden geometrical pattern that is often called "sacred geometry." In 1202 Fibonacci (Leonardo Pisano Bigollo), an Italian mathematician, was influenced by these patterns when he created the mathematical sequence known today as the Fibonacci numbers, or the sequence that each number is the sum of the two preceding ones. For centuries, mystics believed that the divine proportion was the hidden language in nature, and a powerful ratio of healing.[75]

Sacred geometry is seen as spirals, torus shapes, or fractal shapes. Torus and spiral shapes derive from the golden ratio (phi, or $\Phi \approx 1.618$), which is omnipresent in nature, even in the veins of leaves and in the construction of flowers, trees, and mountains. Architecture often uses the principles of the golden ratio, creating structurally sound and aesthetically pleasing buildings. Da Vinci's drawing of a male figure perfectly inscribed in a circle and square, known as the "Vitruvian Man," illustrates what he believed to be a divine connection (the golden ratio) between humans and the universe.[76] In 2019, two neurosurgeons affiliated with the Johns Hopkins University School of Medicine found that the golden ratio is in the architecture of human skulls but is not present in the skull proportions of other mammals.[77]

The golden ratio may be one explanation of why, when we look at something beautiful in nature or in our man-made patterns on structures or materials, we connect with that object—we sense its beauty, if you will. That is one theory of why we resonate with the beauty we see—and perhaps even feel—in nature. The biologist and naturalist E. O. Wilson

The golden ratio, Φ=1+√52, is a sort of magic number, appearing in the most diverse fields of art, biology, mathematics, physics, and engineering, often in connection with fractal growth and self-similarity.

—*Irish Times*

popularized the concept of "biophilia," put forth in his book *Biophilia* (1984), which he describes as "the urge to affiliate with other forms of life." His hypothesis suggests that humans possess an innate tendency to seek connections with nature and that this desire, in part, may be genetic.[78]

There has been a growing body of evidence in the past decade that spending time in nature is good for your physical, emotional, and mental health. Called "Earthing" or "grounding," the idea is that placing your bare feet on the ground, or walking or sitting outdoors in nature, can improve your health and well-being. In Japan, the tradition of *shinrin-yoku*, or forest bathing, is a national pastime believed to reduce stress and promote well-being.

In our highly industrialized society, where roughly 50 percent of the global population lives in an urban area,[79] and the EPA estimates that at least 90 percent of Americans spend their time indoors,[80] we can presumably go weeks without ever touching Earth's surface. This is a frightening statistic given that our bodies sync with the powerful electromagnetic frequencies that nour-ish and guide all the inhabitants of our wonderful home planet. A study published in the February 2013 issue of the *Journal of Alternative and Complementary Medicine* showed that "grounding increases the surface charge on red blood cells and thereby reduces blood viscosity and clumping. Grounding appears to be one of the simplest and yet most profound interventions for helping reduce cardiovascular risk and cardiovascular events."[81] Grounding may also help lower inflammation and reduce stress. According to a study published in the *Journal of Environmental and Public Health*, researchers found that when test subjects were grounded, there was a "rapid activation of the parasympathetic nervous system" (which is sometimes called the rest-and-digest system) "and corresponding deactivation of the sympathetic nervous system" (or the flight-or-fight system, which induces stress).[82]

In other words, touching Earth's surface or hugging—and, yes, we mean really hugging—a tree (like our favorite yogi sister, we at Alice In Futureland take nature walks with, and yes, hug a tree) may

help lower your risk of a heart attack, reduce inflammation, reduce stress, and give you a jolt of Earthly vibrations.

Think about it. Perhaps that "feeling" you get when in nature—whether it's peaceful, or an awakened sense, or a quiet moment of awe—could be a coherence of conversations our body is tuning into, bringing us back to balance and order.

Nature's Fractals Are Contagious

Human stress hormones are reduced by 60 percent simply by observing the fractal patterns in nature. It's an effortless thing that we can simply give ourselves time to do.[83]

Forest Bath

A Japanese concept, *shinrin-yoku* literally means "forest" and "bath" and is about more than just going for a walk in the woods. It involves wandering, using all senses as a guide, to really connect with the natural world. Are we picking up on this supportive and nurturing effort on a cellular level? Studies have found that spending time in wild places such as a forest lowers blood pressure, increases immunity, and elevates mood. And even the New York State Department of Environmental Conservation suggests that you "immerse yourself in a forest for better health."[84]

The Body Light

The Healing Spectrum from the Cosmos to Color

Sense of Coherence

To feel healthy means to feel "coherent." Coherence is a special kind of wholeness in which every part of the body is intercommunicating with every other part. As we learned in chapter 1, our body has intricate electrical networks; we emanate a biofield, and part of that field is our heart field. The research done by HeartMath Institute teaches us that the patterns of our heart activity and emotional states affect our cognitive and emotional function. When you are in a positive state of emotions, your heart's rate is more ordered, sending similarly ordered input to the brain, and you think and perform better.[1] We can experience a sense of coherence, a mixture of optimism combined with a sense of control, as defined by the sociologist Aaron Antonovsky in 1979, to explain why some people stay healthy under stress.[2] We know that when you are in a state of positive emotions such as happiness, it is antithetical, anatomically, to feel unwell. Sounds logical, right?

This interconnectedness from our cells to our emotions is the key to understanding how energy healing works. But before we jump into that resonance, let's start by exploring how the cells in our body share information and keep order.

This energy is from the most powerful source: the Sun. Every creature is engulfed in light that affects its health conditions,[3] and our bodies evolved for millions of years in a field of incoherent light (the Sun), as did all our food sources. Through photosynthesis, plants, algae, and certain bacteria are able to use the energy from sunlight to turn carbon dioxide and water into food. When we eat plants and consume their energy, by extension we too are consuming the energy of those photons—the fundamental particles of light.

The discovery of photons in the body, which we call "biophotons," have shown that particles of the Sun are inside us and facilitate transmission of information among our cells. Biophotons emit light from our cells, an orderly light that helps with cellular communication; and with emerging research, we are beginning to understand what this means to our health. Biophoton emission is

the spontaneous emission of ultraweak light emanating from all living systems, including humans. If you think of the fundamentals of how the human body operates, our intricate biological systems—from circulatory to immune—need to communicate with each other to keep running; cells need to divide and renew to help us repair and grow; and in the end, the human body is one vast information network that fundamentally relies on a constant and dynamic conversation. And that conversation starter is light.

So what exactly is a photon? It is the particle part of light—think of it as the most minute portion of matter. Physicists theorize that photons have no mass or electrical charge.[4] Even so, they are the generators of communication, the sparks that start natural biological and chemical processes (such as photosynthesis) that transmit energy and information. Humans and animals have photoreceptor proteins in our body called cryptochromes,[5] which,

simply put, make us receptive to these photons. These cryptochromes inside us are part of the active components that control our circadian rhythms,[6] the innate clock that regulates when to sleep, when to wake, and repeats roughly every twenty-four hours. Our circadian rhythms are deeply dependent on light, which is why light therapy is used in helping regulate sleep disorders (more on that later).

Photons also have a relationship with proteins such as photolyases, which can help repair damaged DNA.[7] DNA damage can induce aging, cancer, and cell death. Maintaining healthy DNA has been a leading factor to help increase longevity.[8] Photons, the particles of sunlight, are tiny energy and information carriers, helping trigger necessary processes for our health and longevity. Studying biophotons is a relatively new area of research wherein scientists are looking at the presence of light within the body—actually, in our cells.

The nature of the food and the task of the food is not to transport calories. The real task is to order the system, to bring order, to form order, to provide order, to regulate the body. So the food has an obligation more or less to regulate us, to give us a possibility of regulation. And this is caused by photons. We see that the capacity of food, including all other living systems, to submit order, to give the right order to the system—this capacity depends on the storage capacity for light. Look at plants. They are living off light. Not from the caloric point of view. That's not so important. Important is that plant and cells are organized by the sunlight. And this organization which they build up they can transmit to the consumer. So this interaction, evolutionary interaction of order and information, originating originally from the sunlight, from the Sun, and then growing up to systems of higher and higher coherence and information content—this is the real source of that.

—Fritz-Albert Popp, biophysicist, Sputnik Futures interview, 2006

You and I Are Sunshine

Photons within cells were first discovered by Russian scientist Alexander Gurwitsch in 1923, when he observed ultraweak emissions of light while dividing onion cells microscopically. Gurwitsch proposed that living organisms communicate through an exchange of UV light, a phenomenon he called mitogenetic radiation. The radiation he saw was an ultraviolet radiation emitted when the cells were dividing.[9] His concept was rather controversial at the time, until an Austrian physicist, Erwin Schrödinger (yes, the founder of quantum theory), corroborated Gurwitsch's theory when he proposed that living cells can maintain a high level of organization only because the cell's system obtains order from the environment—and this order is provided by sunlight.[10]

In about the mid-1960s the concept of "coherence"—a proposed order in living systems in which light plays a role in delivering this organization—was introduced as a theory by the physicist Herbert Fröhlich. It was a radical theory that became one of the tenets of modern theoretical physics.

But it was in the early 1970s when the theory of biophotons and their role in cellular emission and communication in the cells of plants, animals, and humans was finally verified. Using a highly sensitive photomultiplier technique that can detect and amplify the light from very faint sources, three scientists working independently were able to show the light fields of cells, confirming Gurwitsch's cell radiation theory.[11] One of these scientists was the biophysicist Fritz-Albert Popp, who called this cell radiation "biophotons"—derived from the Greek words *bios* (life) and *phos* (light, power). His research suggests that all living organisms emit a weak, coherent light that generates order and information in the cells to regulate inner processes and actions.[12] Popp is hailed as the inventor of biophoton theory, which proposes that cells communicate with each other inside the organism through biophoton emission, and that DNA may be the most probable source of this biophoton emission.

Let's unpack that: according to Popp, light is constantly being absorbed and remitted by cells, and this light is biophotons. In his work, Popp proposed that the biophotons were being emitted from inside a cell's nucleus by the DNA molecules housed there. These biophotons are constantly firing signals, creating a dynamic, coherent, highly ordered web of light. This web of light is a system that transports information, much like the way fiber-optic wires work, and this web-of-light system could be responsible for chemical reactions within the cells, cellular communication throughout the organism, and the overall regulation of the biological system. In other words, the biophoton emissions act just as powerfully as light does—all you need is a very weak signal or low emission to set off an affect or response. These biophotonic (light) emissions among our cells basically help the body to get organized (coherent).

Popp also discovered that living organisms such as bacteria, sunflowers, fleas, and fish "sucked up" the light emitted by other living organisms in their environment. He concluded that this exchange of photons was a form of communication, a means for living organisms to influence the health of each other. This behavior of organisms using light emissions to send good vibes to their neighborly organisms is a biological and social one, almost like we humans help, treat, and encourage each other toward better health. In ecosystems it is important that all the parts are healthy for the system to thrive and act coherently. In nature, photons encourage coherence and, by extension, promote health. Popp hypothesized that illness results from incoherence, in the form of either too little or too much light.

Biophotons are photons originating from living systems. They are very weak, with intensities of ten to minus seventeen volts, which is very low. It corresponds to a candle in a distance of about five kilometers. And one thinks even if it is so low intensity it is very unimportant. But just the opposite is true. Nature uses so low intensity in order to transmit effectively the information. And by knowing the laws of quantum theory one can speculate that the lower the intensity the higher can be its efficiency in transmitting information. So nature works on a very low level in order to be, in the informational content, very efficient, and in the energetic content with high efficiency also.

—Fritz-Albert Popp, biophysicist, Sputnik Futures interview, 2006

Let's do a little light experiment to understand how we can sense the effects of light. Humor us for a minute and recall a time when you were walking outdoors after hours or days of not being directly in sunlight. Put your book down for a moment, and if you can, walk out and turn your face to the Sun. Or if you are lucky enough to be reading out in sunlight, better still. But if you can't get out or it's not a sunny day, conjure up a memory. Feel (or recall) that warmth from the Sun, even if for just a few minutes. How did it make your feel? Energized? Calm? Happy? Or just really good? That's because the Sun's UV light helps our bodies do a number of things, such as produce vitamin D and set our biological clocks to sleep better. You probably don't feel that happening, but that sunlight-good feeling is a reminder that all those rays are helping your emotional wellness. Exposure to sunlight boosts the feel-good chemical in your brain called serotonin, which helps you with energy and keeping calm and focused. Too little exposure to light can make us

Risk Factors of Seasonal Affective Disorder

Being female: SAD is diagnosed four times more often in women than in men.

Living far from the equator: SAD is more frequent in people who live far north or south of the equator. For example, 9 percent of those who live in New England or Alaska suffer from SAD compared to just 1 percent of those who live in Florida.

Family history: People with a family history of other types of depression are more likely to develop SAD than people who do not have a family history of depression.

Having depression or bipolar disorder: The symptoms of depression may worsen with the seasons if you have one of these conditions (but SAD is diagnosed only if seasonal depressions are the most common).

Younger age: Younger adults have a higher risk of SAD than older adults. SAD has been reported even in children and teens.

—National Institute of Mental Health[13]

not feel our best, maybe even off our game. Doctors now recognize that lack of light due to seasonal changes is a type of depression known as seasonal affective disorder (SAD), typically starting in the late fall and early winter and often going away during the spring or summer. According to the National Institute of Mental Health, the root causes of SAD are still not confirmed, but one of the treatments is natural or artificial light therapy.[14]

Light Is "Woven into the Human Fabric of Healing"

So we had a little sunlight break (and hope you tried it!). We learned that there exists communication among the cells of all living plants, animals, and people, and that this form of communication is caused by natural light through a phenomenon called "biophotons." We also learned that these biophotons are carriers of "information," without which our bodies may just be lifeless collections of molecules. According to Popp, the biophotons within the body mainly originate in our DNA. All living things have DNA within their cells. It's the building blocks that give us structure and function. Human beings, and all living things on Earth, are open systems—we take in energy from our environment, and this energy includes light. Light is necessary in bringing order and coherence. And that coherence is what supports homeostasis and ultimately health.

But just how does that ray of UV light activate our hormones, or tell our bodies when to sleep?

Light Therapy Treatment for SAD

Light therapy has been a mainstay of treatment for SAD since the 1980s. The idea behind light therapy is to replace the diminished sunshine of the fall and winter months using daily exposure to bright, artificial light. Symptoms of SAD may be relieved by sitting in front of a light box first thing in the morning, daily from the early fall until spring. Most typically, light boxes filter out the ultraviolet rays and require twenty to sixty minutes of exposure to ten thousand lux of cool white fluorescent light, about twenty times greater than ordinary indoor lighting.

—National Institute of Mental Health[15]

Well, without getting too far into the weeds, scientists believe that our cells actually absorb the photons, and use this light to transmit information to each other, and that gets the whole system to stay organized and maintain a steady state.

The Wellman Center for Photomedicine at Massachusetts General Hospital is an academic research facility dedicated to investigating the effects of light on human biology and the development of light-mediated, minimally invasive diagnostic and therapeutic technologies. In their article "Why Light for Medicine?" they explain the very cosmic way by which photons that originate from the Sun and other stars reach us in the form of waves, and these photons stimulate our cells. For instance, when you gaze at stars in the night sky, you are activating the biochemistry in the eye's retina via these photons. Just as astronomers built their knowledge of stars by reading the photons they emit, the photons in our "inner space" (our body) can, in a similar way, tell us what's going on inside ourselves. The article posits that light is "woven into the human fabric of healing" and that light therapy may be the way to directly change what's going on inside our molecules.[16]

The Light Contagion

Living things communicate with each other through photon emissions. Plants can make other plants grow. Animals can induce sickness in others nearby, not through physical touch or contagion, but simply through photon radiation. Healers impact others through this same energy and information transfer.

—Buddhist Door[17]

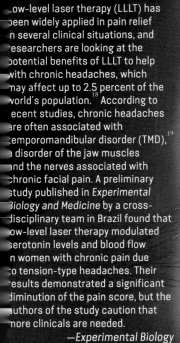

Low-level laser therapy (LLLT) has been widely applied in pain relief in several clinical situations, and researchers are looking at the potential benefits of LLLT to help with chronic headaches, which may affect up to 2.5 percent of the world's population.[18] According to recent studies, chronic headaches are often associated with temporomandibular disorder (TMD),[19] a disorder of the jaw muscles and the nerves associated with chronic facial pain. A preliminary study published in *Experimental Biology and Medicine* by a cross-disciplinary team in Brazil found that low-level laser therapy modulated serotonin levels and blood flow in women with chronic pain due to tension-type headaches. Their results demonstrated a significant diminution of the pain score, but the authors of the study caution that more clinicals are needed.

—*Experimental Biology and Medicine*[20]

Healing Beams

In modern light therapy, there typically are two forms of light used: coherent and incoherent. To understand the difference between coherent and incoherent light, think of a flashlight. If you point it to a wall in a dark room, the beam it projects is softly diffused and spread out a bit—it is not a perfect, intense circle of light. If you aim a laser pointer or a laser therapy tool in the same way as you did the flashlight, you will see a tiny spot on the wall because the laser beam remains coherent across the distance.[21]

Most visible light on Earth and in the universe, including the Sun, light from incandescent fluorescent, and LED sources, is incoherent. We use this form of light for most home light therapies, also known as bright light therapy or phototherapy, to treat SAD, as we explained earlier, as well as depression, mood, and sleep disorders.[22] Light therapy involves a light box, lamp, or other type of light device that mimics natural outdoor light by giving off a bright light.

Lasers, on the other hand, are coherent light forms. With lasers,

light can stay tightly focused without spreading out, making laser therapy a medical treatment ideal for procedures that require light to cut, burn, or destroy tissue. The term "laser" stands for "light amplification by stimulated emission of radiation." Lasers are used for many medical treatments and surgical procedures such as treating varicose veins, improving vision, removing kidney stones and tumors, and during some skin surgeries.[23]

One form of light on the spectrum that has been popular in healing skin disorders is UV light, which helps in treating people who suffer from moderate to severe psoriasis. Other benefits include reducing inflammation and slowing the production of skin cells. Because the radiation from UV light can be intense, the treatment is usually given in a dermatology practice.[24]

LED sources are unique in that they emit a narrow spectrum of light in a noncoherent manner. LED light devices using varying wavelengths of red, blue, and green are commonly used for skin care treatment, from handheld,

What I'd like to start seeing us do is beginning to understand issues about what parts of the eye intercept which wavelengths. One of my doctoral students was working on circadian rhythm and light, her interest was in circadian rhythm, and it turns out there's a particular wavelength of blue light that needs to enter in the periphery of your eye to set your circadian rhythm. Something that happens automatically when you're outside because you've got a lot of scattered blue light. Something that doesn't happen automatically on the inside. She was beginning to look at how does she pull out that wavelength or find that wavelength. Perhaps it's coming in from a little bit of Sun, a way of extracting and getting it to the side of the eye. Her dissertation was very much about this issue of extracting wavelengths. We wouldn't notice that there was actually a blue wavelength entering the eye in a particular way, yet it would have profound consequences on how somebody felt.

—Michelle Addington, architect, engineer, Yale Graduate School of Design, Sputnik Futures interview, 2007

wand-like devices to Jason-esque full-facial masks (yes, that Jason from the Friday the 13th movies!) that use varying wavelengths of light, including red and blue. But here's a fun fact: this prestige light treatment for skin was originally developed by NASA for plant-growth experiments on shuttle missions. A team of doctors from the Medical College of Wisconsin in Milwaukee used the technology in 2000 to examine how LEDs could help to heal hard-to-heal wounds such as diabetic skin ulcers, serious burns, and severe oral sores caused by chemotherapy or radiation. They found that the near-infrared light emitted by LEDs seemed to help increase energy inside cells. According to Dr. Harry Whelan, professor of pediatric neurology and director of hyperbaric medicine at the Medical College of Wisconsin, "This means whether you're on Earth in a hospital, working in a submarine under the sea or on your way to Mars inside a spaceship, the LEDs boost energy to the cells and accelerate healing."[25]

The study of the beneficial and harmful effects of light in living organisms is called photobiology. Global culture is in an interesting

age where biology is becoming our technology, what some call "transformational biology," that is impacting medicine and new environmental sustainability solutions. Light is one of its most powerful tools that is just beginning to be tapped.

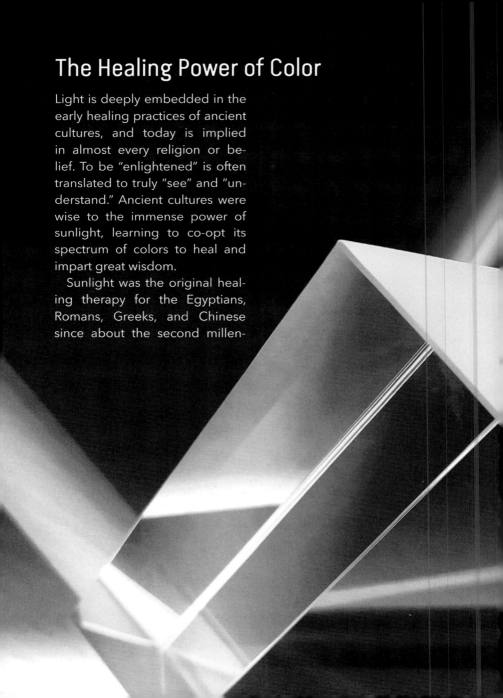

The Healing Power of Color

Light is deeply embedded in the early healing practices of ancient cultures, and today is implied in almost every religion or belief. To be "enlightened" is often translated to truly "see" and "understand." Ancient cultures were wise to the immense power of sunlight, learning to co-opt its spectrum of colors to heal and impart great wisdom.

Sunlight was the original healing therapy for the Egyptians, Romans, Greeks, and Chinese since about the second millennium BC. A common technique across these early cultures was using sunlight to create the "color light" that worked its healing magic. Some Egyptian temples were used for healing ceremonies, using the rays of sunlight to shine through colored gems such as rubies and sapphires, which were strategically placed on people's bodies (the possible precursor to the alternative method of crystal healing, the use of semiprecious stones and crystals to promote the flow of good energy and help remove bad energy). The Egyptians were also said to construct rooms that radiated a particular color from sunlight. But it may be the ancient Greeks who first used healing temples with specific spectral energy of sunlight and colors, documenting the idea of "solar therapy." In Heliopolis, the Greek City of the Sun, these healing temples used a specific color of light for a certain medical problem.

To the Chinese, color is regarded as a part of the cosmic energy qi, the vital life force that can shape energy and one's des-tiny. From the second millennium BC, the Chinese used color to indicate directions for north (the color black), south (red), east (green/blue), and west (white). They also had colors for identifying seasons, the cyclical passage of time, and the internal organs of the human body.[26]

The idea that the human body is stimulated by colors, and that colors can work to help various systems in the body to function, is an alternative or complementary medical practice known as chromotherapy. Also called color therapy, chromotherapy is described as a "method of treatment that uses the visible spectrum (colors) of electromagnetic radiation to cure diseases."[27] It draws on the roots of centuries-old Ayurvedic medicine and the ancient Egyptian healing practice of Sekhem, which combines spiritual and mystical teachings, as well as traditional Chinese healing. The practice is becoming more popular in medical spa treatments and in alternative and complementary practices.

Life Is Meant to Be Lived in Color

What Is Color?

Light comes to Earth from the Sun in waves. Some of the waves are longer. Some of the waves are shorter. We see the waves as the colors of the rainbow. Each color has a different wavelength. Red has the longest wavelength. Violet has the shortest wavelength. When all the waves are seen together, they make white light.

We think that white has no color, but that is not true. One way to see this is to shine white light through water. When white light shines through water, it makes a rainbow. The water breaks apart the color of white so we see all the colors of the rainbow. We call these colors the visible light spectrum. Our eyes have cones inside them to help us see the waves as color. Our eyes see the reflection of sunlight off things around us. The color we see is the light reflected back to our eyes.

—NASA[28]

How Do We See Color?

The human eye sees color over wavelengths ranging roughly from 400 nanometers (violet) to 700 nanometers (red). Light from 400 to 700 nanometers (nm) is called visible light, or the visible spectrum. Light outside this range may be visible to other organisms but cannot be perceived by the human eye. Colors of light that correspond to narrow wavelength bands (monochromatic light) are the pure spectral colors learned using the ROYGBIV acronym: red, orange, yellow, green, blue, indigo, and violet.

Some people can see farther into the ultraviolet and infrared ranges than others, so the "visible light" edges of red and violet are not well defined. Also, seeing well into one end of the spectrum doesn't necessarily mean you can see well into the other end of the spectrum. You can test yourself using a prism and a sheet of paper. Shine a bright white light through the prism to produce a rainbow on the paper. Mark the edges and compare the size of your rainbow with that of others.

—Anne Marie Helmenstine, PhD, ThoughtCo[29]

Chromotherapy, Color Me Well!

A paper published by Samina T. Yousuf Azeemi and S. Mohsin Raza in *Evidence-Based Complementary and Alternative Medicine* poses that several studies and published material have elaborated on the relationship between the body and colors, and that these various materials provide documented systems of treatment using the so-called characteristics of color.[30] The researchers propose a different research methodology to assess how color therapy works, illustrating their recommendation to look at the development of science in the field of electromagnetic radiation/energy as a comparison.

Their rationale: first, we know that light is energy, and that color is a product of the wavelengths of light. A specific wavelength, a certain frequency, and a particular amount of energy in that wave dominate to make it a dis-

tinct color. But they add in one more crucial point: light is electromagnetic radiation, which is the fluctuation of electric and magnetic fields in nature. Azeemi and Raza go on to say that "contemporary medicine examines the symptoms and influences or suppresses them, but it does not involve itself with a real source—diseased life energies." This means that in every organ, for example, there is an energetic level at which the organ functions best. That is mostly due to the fact that our cells are specialized to conduct electrical currents (more on this in chapter 4!), and electricity is, of course, a form of energy. Therefore, when a part of our body deviates from its normal vibrations, it could imply that the part of the body is either diseased or at least not functioning properly. The authors reason that colors in chromotherapy treatment generate electrical impulses that are "prime activators of the biochemical and hormonal processes in the human body, the stimulants or sedatives necessary to balance the entire system and its organs."[31]

Azeemi and Raza trace the history of chromotherapy in their comprehensive paper, citing doctors and scientists who have recognized color and its effect on the body, but caution that the established fields of medicine today still don't recognize chromotherapy as a valid scientific treatment method. Still, there is a time-honored tradition of thinking about the intersection of health and colors. In 1025, Avicenna, a Persian polymath and one of the most significant physicians and thinkers of the Islamic Golden Age, wrote that "color is an observable symptom of disease" in his eminent encyclopedia *The Canon of Medicine*. Avicenna also developed a chart that related color to the temperature and physical condition of the body. His view was that red moved the blood, blue or white cooled it, and yellow reduced muscular pain and inflammation. Many scientists followed suit in looking at color, light, and healing. Physician Edwin D. Babbitt (1828–1905) was one of the early proponents of chromotherapy, conducting experiments on the soothing and stimulating properties of color, which he outlined in his book *The Principles of Light and Color: The Classic Study of the Healing Power of Color* (1878). Babbitt identified red as a stimulant affecting blood flow, and blue and violet as soothing and having anti-inflammatory properties.[32]

More recently, Charles Klotsche, a health practitioner and author of *Color Medicine: The Secrets of Color Vibrational Healing*, discusses using chromotherapy as a complete therapeutic system for 123 major illnesses, using single colors and also combinations of two or more colors for therapy techniques. In his book, Klotsche emphasizes his support for chromotherapy as safe, simple, economical, and highly effective.[33] This may be one of the reasons why chromotherapy is sought as an alternative, noninvasive way to help those suffering from pain, depression, anxiety, lack of energy, sleep disorders, and more. Curious about the effects of different colors of light on sleep, a team from the Sleep and Circadian Neuroscience Institute at the

University of Oxford conducted an experiment where they exposed mice to three different colors of light: violet, blue, and green. They found that while green-light exposure produced rapid sleep onset—between one and three minutes—when exposed to blue and violet light, the onset of sleep was delayed, taking between sixteen and nineteen minutes for blue and between five and ten minutes for violet.

The study also looked at the regulation of the level of corticosterone stress hormone during sleep to better understand the effects of different colors of light on regulating the hormone. Their findings suggest that there are "different pathways in the brain, by which different colors of light have different effects on sleep or wakefulness," and concur with other recent studies showing that the blue light from these devices delays sleep.[34] The researchers concluded that the results add to our understanding of the effects of light-emitting devices on humans, particularly the blue screens of our mobile phones, which many people use before going to bed. Still, they are careful to assert that "we need to understand how the overall color balance of artificial light could affect people's alertness and sleep."[35] As more research emerges on the biological effects of colored light, perhaps chromotherapy options for your mobile phone could become the next big app akin to the popular meditation apps today.

The Healing Spectrum

Chromotherapy is said to align our major organs and limbs of the body with one of the colors in the spectrum (red, orange, yellow, green, blue, indigo, and violet), using a distinct color that resonates with that body part. The underlying theory is that our organs, cells, and atoms contain energy, and each has its own form of frequency or vibrational energy. The colors of the light spectrum are actually wavelengths of sunlight. For example, red has the longest wavelength. Violet has the shortest wavelength. (Wavelengths are the distance between crests of an electromagnetic or sound wave, which are frequency.) In chromotherapy, the color red is used to support circulatory and nerve functions. Another example is the use of blue light in the treatment of a variety of psychological problems, including addictions, eating disorders, and depression. Some spas and well-being centers offer a choice of color frequencies to be beamed into the infrared sauna, prescribed as helping to balance the body's energy and promote better emotional, physical, and mental well-being.[36]

RED

Chromotherapy is used to improve the performance of athletes: red light appears to help athletes who need short, quick bursts of energy, while blue light assists in performances requiring a steadier energy output.

—*Evidence-Based Complementary and Alternative Medicine*[38]

PINK

Strong pink acts as a cleanser, strengthening veins and arteries. Pink activates and eliminates impurities in the bloodstream. Pink light has a tranquilizing and calming effect within minutes of exposure. Pink holding cells are now widely used to reduce violent and aggressive behavior among prisoners.

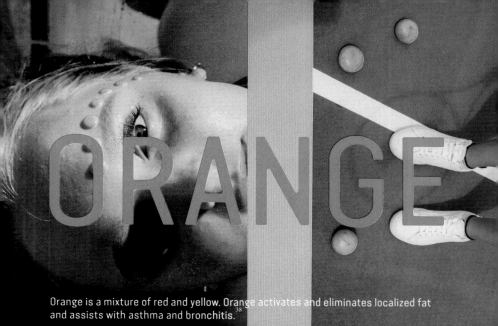

ORANGE

Orange is a mixture of red and yellow. Orange activates and eliminates localized fat and assists with asthma and bronchitis.[38]

YELLOW

Strong yellow strengthens the body and activates internal tissues. Yellow, the brightest color used in chromotherapy, has been used to purify the skin; help with indigestion; strengthen the nervous system; treat glandular diseases, hepatitis, and lymphatic disorders; and assist metabolism.

GREEN

Strong green provides anti-infectious, antiseptic, and regenerative stimulation. Green provides a neutral, positive, calming effect.[40] Green light may be an alternative treatment for migraines according to researchers at Harvard Medical School's Beth Israel Deaconess Medical Center, who found that exposing migraine sufferers to a narrow band of green light helped reduce headache severity.[41]

BLUE

Strong blue lubricates joints and helps address stress, nervous tension, and infections. Blue exhibits tranquilizing qualities often used to relieve headaches, colds, stress, nervous tension, rheumatism, stomach pains, muscle cramps, and liver disorders. Blue is thought to have a positive effect on all kinds of pain.[42]

INDIGO

Indigo is used to address conditions involving the eyes, ears, and nose. It has a calming, sedative effect.[43]

VIOLET

Violet is used to calm the nervous system, soothe organs, and relax muscles. Violet has meditative qualities and is often used to treat conditions of the lymphatic system and spleen as well as urinary disorders and psychoses.[44]

The Chakra Colors

We can't talk about colors and frequency and healing without touching on probably one of the most widely known yet totally ancient forms of energy: chakras. (Fellow yogis, you may know this already, so if you would rather jump ahead to the next chapter, be our guest!)

The word "chakra" derives from Sanskrit as far back as 1500 BC. Interpreted to mean a wheel or a circle, the concept of the chakra is thought to first be a metaphor for the Sun. Chakras embody the seven energy centers within the human body, and each is said to relate to a person's physical, mental, emotional, and spiritual self. According to Yoga International, the chakras are in "locations where subtle energy channels, known as nadis, converge."[45] These converging hubs align with the spine, starting from the base of the spine (near your pelvic area) and following up to the crown of the head, the "soft spot" area on top of our skull we are born with but that eventually closes at about two months of age. One way to visualize chakras is to think of spinning pinwheels, colorful and spiraling when moving with the energy of the wind. That energy wind inside of us is referred to as prana (breath), the vital life force in Sanskrit.

There has been much interpretation of the original ancient scripts that spoke of many chakra systems throughout the body, and the yoga practices adopted by Western society, mainly using the seven-chakra system that originated in a 1577 Sanskrit text by Pūrṇānanda Yati. The psychological states associated with each chakra started with the teachings of Carl Jung. The charting of colors to represent a specific chakra has a twentieth-century origin, most notably through the research of Dinshah Ghadiali (1873-1966), a researcher and expert in radioactivity who discovered that the body could be adjusted from disease to health by systematically exposing it to colored light. In his *Spectro-Chrome Metry Encyclopaedia*, Ghadiali explained that particular areas of the body respond to particular colors, and that these areas are similar to what the ancients called chakras. Ghadiali presented chakras as the source of energies.[46] Each of the chakras have an associated color based on that color's frequency, and the colors are the spectrum of light we have been discussing, which are also the colors of a rainbow.

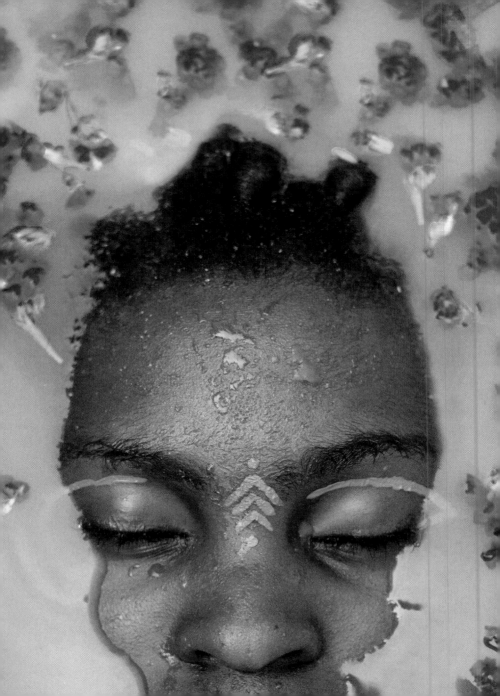

The Seven Chakras + Colors

First chakra, or the root chakra, is at the base of your spine near your tailbone. This chakra represents your "power base"—security, stability, and basic needs. Its color: red. (And remember, red has the longest wavelength.)

The second chakra is the sacral chakra, associated with orange. It is about two inches below your navel and is said to generally govern your creativity, sexuality, and emotional well-being.

The third chakra is the solar plexus chakra, and—you guessed it!—its color is yellow. It's yellow because this chakra, between our navel and sternum, the lower part of the breastbone, is all about power, confidence, and self-esteem, characteristics associated with the omnipotent Sun.

The fourth chakra is the heart chakra, which you might think would be red or pink, but its hue is actually green—for empathy, compassion, kindness, healing, health, and yes, love. Your heart chakra lives in the center of your chest and is considered to unite the lower chakras (viewed as the chakras of matter) with the upper chakras (the spirit chakras), just as green marks the intersection of warm and cool colors in the light spectrum.

The fifth chakra represents communication, self-expression, and speaking your truth. It is the throat chakra, at the center of the throat, and its color is blue.

The sixth chakra is the third-eye chakra, at the center of your forehead between your eyebrows. It is influenced by the color purple, which in some theologies is considered the mystical and magical color of wisdom. The third-eye chakra is about connections to intuition, inner vision, and consciousness.

The seventh chakra is the crown chakra, at the top of the head, and some say extends about two inches above your crown. This is because the crown chakra is the enlightened one, connecting you to your higher self, the universe, and the divine. Its color is violet, the shortest wavelength in the color spectrum. Human eyes are not sensitive to short wavelengths, which may be why it represents the spiritual and not the physical part of us.

—Well+Good[47]

In the practices of yoga, meditation, and Ayurveda, the idea is
to use your breath to help keep
the energy open in each chakra.
Practitioners are encouraged to
concentrate on one chakra area
at a time, inhaling through the di-
aphragm so the belly pushes out,
really filling your lungs, usually to
a count of four or more, and then
exhaling, drawing the breath out
from the abdomen. While there
are a few variations to the breathing and visualization techniques
used in mindful practices, one
key is to envision the chakra you
are sending energy to by also imagining the corresponding color.
Try it for a minute. You don't have
to be in the lotus position; just get
yourself quiet and focused on a
particular chakra, starting with the
root; inhale one-two-three-four,
visualize the color red. And now
exhale, four-three-two-one. Do a
few more, and color light yourself
with some prana.

Can Life Feel Light?

Color and light are deeply rooted in the early healing practices of ancient cultures. Light defines our universe, and sunlight is the primary source of energy for all life on Earth. Our bodies thrive on incoherent light from the Sun, and our cells emit light (biophotons). The DNA in our cells can produce coherent light, carrying the precise information required for the growth, functioning, and healing of our magnificent bodies. But what is more interesting is that the photons absorbed by our retinas were created in the Sun's core tens of thousands of years ago before being emitted by the Sun. But once that photon escaped the Sun's surface, it took only eight short minutes to arrive on Earth—and make its way inside you.[48] Yes, you are the body light!

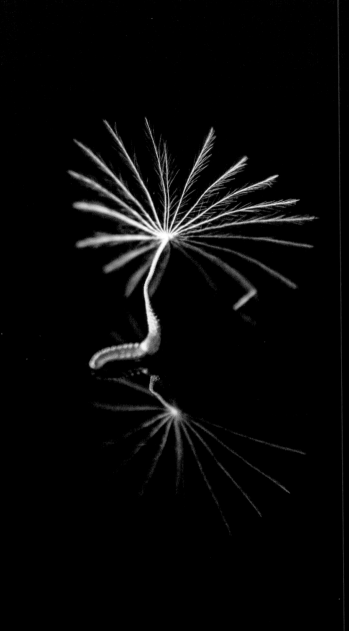

Frequency Healing

Turn On, Tune In, and Take
Your Electromedicine

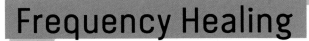

Our journey so far has looked at making the invisible visible, from the electromagnetic energy fields of the Earth to those in nature to the energy fields in our bodies.

In chapter 2 we dropped in on nature and learned how everything from animals to plants are tuned into electromagnetic fields, using them to guide growth and navigation. With our reading minds, we listened in on the harmonic songs of conversation, from whales to birds, and discovered that our cells talk to each other by emitting light—and that light originates from the Sun. And we saw how the Sun's rays create a colorful spectrum that heals us inside and out. Ah, it truly is a universe of beautiful vibrations . . . and one that is bioelectric. New science is capitalizing on these discoveries, manipulating and co-opting the behavior of electrical pulses to create a new field of medicine that may one day make the notion of pills or drugs obsolete.

The key to understanding vibrational healing lies in rerouting energy fields that form complex relationships with other fields, such as those surrounding the physical/cellular substance and others relating to more nonphysical energies.

—Charles Klotsche, *Color Medicine*[1]

Biology 101 has taught us that the body has an electrical system that helps our hearts beat, our muscles twitch and respond, and our body communicate with our brain. When Fritz-Albert Popp first discovered that all living cells emit light as biophotons, he could not have anticipated the revolution this would create in biology and physics. Since then, there has been a great wave of research in the pioneering fields of biophotonic diagnosis, biophysics, biofields, and biomagnetism. Popp is one of possibly hundreds of brilliant scientists and researchers who continue to fearlessly investigate the potential of electrical dynamics in the human body. The notions that the body is electric, that light is the way our cells communicate, and that we have an alignment of energetic pathways are both ancient and contemporary ideas that are the premise of vibrational medicine, or vibrational healing, as we covered in the introduction and chapter 1. Modern science is taking this all one step further—or, should we say, to a higher frequency—with the emerging field of "bioelectric medicine" or "electroceuticals."

The Bioelectric Switch

We are entering a dynamic new era of bioelectric medicine, which leverages the convergence of neuroscience, electronics, materials science, molecular medicine, and biomedical engineering to create devices and ways to treat chronic diseases. The goal of bioelectronic medicine is to restore healthy patterns of electrical impulses—for example, adjusting how neurons fire and neurotransmitters travel through our neural circuits. Within the next decade, modulating the body's neural networks could become a mainstream therapy for many of today's greatest health issues—from arthritis,[2] asthma,[3] and Alzheimer's disease[4] to depression,[5] diabetes,[6] and digestive disorders.[7] According to studies, stimulating nerves also shows promise in treating cardiovascular disease[8] and even in improving cognition.[9, 10]

A growing method of bioelectric stimulation is neuromodulation, which alters nerve activity through targeted delivery of a stimulus, such as electrical stimulation, to specific neurological sites in the body. The core premise is to stimulate or reduce the activity of specific nerves with either external sensors or implanted devices that can be activated when needed. It's not the way Dr. Victor Frankenstein experimented with electricity to bring a dead man back to life in Mary Shelley's classic 1818 novel. Neuromodulation is a more precise and controlled electrical stimulation sent to specific nerves that are meant to target cells in an organ, and the expected outcome is to stimulate or regulate the body's immune and metabolic responses.

Bioelectric medicine pioneer Dr. Kevin Tracey and his team at the Feinstein Institutes for Medical Research in New York have been experimenting with silicon wafers made of superthin film electrodes that can be implanted into a patient to record from or stimulate the nervous system, targeting the vagus nerve. An article published in *Proceedings of the National Academy of Sciences* (PNAS) tells the story of how Tracey and his team helped alleviate a patient's inflammatory arthritis and Crohn's disease, which had crippled her for most of her early adult years. In 2017, a research team implanted a small device inside her chest to stimulate her vagus nerve—and three years later, she is in clinical remission. If you recall our discussion on the vagus nerve and emotions in chapter 1, the vagus nerve is the longest nerve in the body, linking the brain to organs from the esophagus to the intestines while controlling breathing and heart rate. It is ideal for neuromodulation, as it carries about a hundred thousand nerve fibers through the body.[11]

The vagus nerve is also linked to the brain-gut axis and is an important modulator in treating psychiatric and inflammatory disorders.[12]

Conditions such as headaches, chronic pain, and even opioid addiction may also be alleviated through some form of neuromodulation. So far, the health-care industry has been testing the effects of neuromodulation with a therapy that involves a spinal cord stimulator that when surgically implanted and turned on sends mild electric pulses to the nerve fibers in the spinal cord. The effect is like a disruption; the low-level electricity interrupts the path of the pain signals that are carried to the brain, providing some relief to chronic pain sufferers.

Of course, nerve stimulators have been used to treat conditions such as epilepsy, mental illness, and depression for decades, and forms of electrical stimulation to the body and tapping into the body's electrical network go back centuries. One of the godfathers of bioelectrical medicine is Luigi Galvani, professor of anatomy at the University of Bologna in Italy, who in 1780 recorded studies of the effects of electricity on muscular movement in frog's legs. In one experiment, he touched a frog's legs with a pair of scissors during an electrical storm, observing a twitching movement.[13] A century later, two great inventors whom we covered in chapter 1 were experimenting with their understanding of electromagnetism on the human body. The inventor Michael Faraday is credited with discovering electrolysis, a technique that uses a direct but weak electrical current to drive a reaction. And in the United States, Nikola Tesla experimented in about 1891 with the use of frequency in heating deep body tissues.

Scientists have expanded on these original studies, and today we know that electrical stimulation activates the same pathways of the electrical signaling that happens naturally in our bodies. These electrical signals regulate many of our cellular functions, such as hormone production and muscle contraction, and pass through nerves between the brain and the organs where the cells are located. The frequencies of these currents determine how active the cells are in performing their assigned functions.[14] Bioelectric therapy makes that frequency strong enough to help the cells perform.

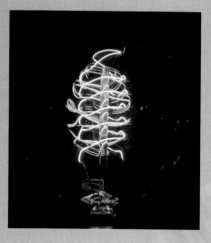

Neurostimulation to the Rescue

With the advances today in nano-thin electronics with small-scale sensors and stimulators that are biocompatible and can be implanted in the body, we have the potential for new electricity-based therapies to help with everything from chronic pain to high blood pressure, arthritis, diabetes, and even dementia.[15] These new treatments delivered in clinic and at home will disrupt the prescription drug industry—and this is just the beginning.

Children's Wisconsin in Milwaukee is one of the few centers in the United States that is using a form of bioelectronic therapy called auricular neurostimulation, a non-pharmacological alternative to help treat functional gastrointestinal (GI) disorders such as irritable bowel syndrome in children. The treatment is a small wearable device with a battery and four electrodes that is attached to the ear with adhesives. The electrodes pierce the skin to deliver low-voltage currents (3.2 V) with alternating frequencies that the hospital claims most patients do not even feel. These microcurrents are said to stimulate branches of several cranial nerves, possibly including the vagus nerve, in the ear.[16] In June 2019, the FDA approved auricular neurostimulation for the treatment of irritable bowel syndrome based on the study performed at Children's Wisconsin.[17]

Abbott Laboratories, a global health-care company that makes products for diagnostics as well as medical devices, nutritionals, and medicines, has a division dedicated to neuromodulation devices for movement disorders and for treating chronic pain—both of which were usually treated by chemical stimulants, aka drugs.

Two things a prescription drug can't help with (just yet) are memory and dementia, a growing focus in new neuromodulation studies. Scientists at Boston University recently published a study that showed they were able to improve the working memory in adults ages sixty to seventy-six using a harmless form of electrical brain stimulation. Think of your working memory as your immediate mental calculator. It's the part of the brain system that

Wearable Migraine Relief

Cefaly is a nondrug migraine treatment device that uses precise microimpulses sent through a device that you can attach to your forehead using a self-adhesive electrode. FDA-approved under prescription, the Cefaly device has two settings for the micropulses that are sent to the upper branch of the trigeminal nerve. The Acute setting helps to relieve the headache pain during a migraine attack. The Prevent setting is used to prevent future migraine attacks.

—Cefaly

Neurotechnology

Quell is a 100 percent drug-free wearable technology for pain relief, using neurotechnology for precise and personalized management of chronic pain. The Quell stimulator slips into a wraparound band, and users apply electrodes with hydrogel pads and silver contacts directly onto the skin. Quell stimulates nerves over an extended period, controlled by a companion app that lets the user calibrate the level of stimulation and customize the duration of the therapy. The app acts like a coach tracking your patterns and also alerts you to weather changes that may impact your pain.

—Quell

holds information for short periods, and is used to make split decisions, recognize a familiar face, or do simple arithmetic. Working memory is known to decline with age, even if you don't have the onset of dementia. The culprit in this form of memory decline may be that the electrical brain sync between the prefrontal and temporal regions is slowing down. In this study, the control group of older adults showed improved working memory after the electrical intervention, and the effect appeared to last for fifty minutes after the stimulation.[18]

An App for Your Pain

Consumer health electronics are getting into the game of neurostimulation therapies with wearables and home devices focused on alleviating chronic pain through microcurrents that gently stimulate the nerves and nerve endings of muscles. Many come with an app to set the intensity and duration of the electrotreatment.

Smartphone-Controlled Migraine Relief

Instead of placing electrodes and other cumbersome devices on the head, Nerivio is a remote electrical neuromodulation (REN) arm-strapped wearable patch to treat migraines. Nerivio patches contain electrodes, a battery, and a proprietary smart chip that syncs with the Nerivio app, which controls the patch, turning it on, selecting the intensity and duration.

—Theranica

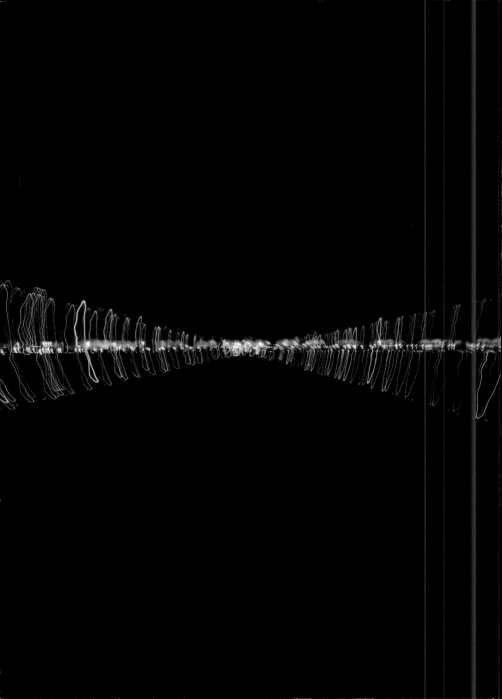

Zapping Is the New Pill Popping

The fields of nanotechnology, biology, and radio physics are creating a new kind of "drug"—call it electroceuticals or electromagnetic pharmacology—but whatever the label, researchers are increasingly tuning in, and the implications are huge.

Although bioelectronic medicine is in its early days, two major players—British drugmaker GlaxoSmithKline and Verily Life Sciences, a division of US tech firm Alphabet (parent of Google)—have made a collaboration focused on developing electricity-based therapies. In 2018, researchers at Massachusetts Institute of Technology, Draper Laboratory, and Brigham and Women's Hospital designed an electronic pill—basically an ingestible capsule—that can be controlled using Bluetooth wireless technology. The electronic capsule can relay diagnostic information or release drugs, responding to commands from a user's smartphone, and could stay safely in a person's stomach for at least a month before it breaks down and passes through the digestive tract.[19]

Implants that are biodegradable, in that they can disappear after they are done doing their job inside you, are not far off. John Rogers, a materials scientist at Northwestern University, and his colleagues have developed a biodegradable implant that can continuously deliver electrical pulses to nerves, and then break down when it is no longer needed. The device is about as thick as a piece of paper and roughly the width of a dime and is flexible enough to wrap around an injured nerve. The device is powered and controlled wirelessly by a transmitter outside the body and electrically stimulates the nerve for about two weeks before the body absorbs it. It is fairly common practice during operations on patients who have nerve damage to apply some electrical stimulation during the surgery to aid in recovery. The research team at Northwestern hopes that one day this digestible device will continuously provide that added boost of electrical stimulation at various points throughout the recovery and healing process.[20]

Electrified Meds

In her 2019 article "Why It's Time to Take Electrified Medicine Seriously" for *Time*, the staff writer and Center for Social and Economic Research media fellow Alice Park explains the ways scientists anticipate the potential of electrical medical solutions, including the possibility that patients with rheumatoid arthritis will be able to "turn on an implanted electrical device to quiet the immune response that drives their painful inflammation." Electrical signals are being tested for things such as activating nerves in the eye to restore vision (Massachusetts General Hospital) and manipulating electrical signals in the brain to address conditions from depression to dementia (Johns Hopkins).[21]

Electrical stimulation of the brain, known as transcranial electrical stimulation (tES), also has a long history. It was first used in Victorian and Edwardian times in the form of electroshock therapy to elevate mood and improve mental performance,[22] though of course most of those outcomes have not been proven. Today, researchers are looking for a more precise tool to control neurons in the brain without stimulating or damaging nearby cells; this is a challenge for neuroscientists, especially in researching alternative treatments for Parkinson's disease. One solution that is gaining traction is optogenetics, which came onto the scene in 2005. Optogenetics uses a combination of light and genetic engineering to control the cells of the brain, possibly turning neurons on and off to alleviate the movement-related symptoms of Parkinson's disease. Parkinson's disease is a disorder of the central nervous system that causes tremors and eventually impacts movement. It is generally caused by nerve-cell damage in the brain. Today, electrodes clinically inserted into the brain can be turned on and off to help with these tremors. But by activating the cells in the basal ganglia (midbrain) area using optogenetics, researchers believe they can relieve symptoms for much longer than current deep-brain stimulation therapies or pharmaceuticals.[23] Researchers are testing how optogenetic LEDs implanted into the brain can directly

stimulate the neurons affected by Parkinson's disease without sending any electrical impulses to the neighboring neurons.[24]

The promise of nonevasive, drug-free, and wirelessly controlled treatments has financial analysts projecting the global bioelectric medicine and electroceuticals market to reach $35.5 billion by 2025.
        —Research and Markets[25]

Electrical Skin

John Rogers is a pioneer in the new field of wearable, stretchy, and dissolvent electronics for the body. In 2011 he and his colleagues at the University of Illinois at Urbana-Champaign, where Rogers was then head of the Department of Materials Science and Engineering, published a paper on their development of an "epidermal (skin) electronic system," an ultrathin electronic device that stuck to the skin without adhesives or gels. Applied like a temporary tattoo, the device was held in place on the skin by the weak van der Waals electrical forces that exist between molecules of the same substance. These weak forces held the device in place without interfering with normal skin motion, meaning that the tattooed device could bend, scrunch, and stretch without damaging itself or the skin.

The team tested the "electronic skin" device to measure heart rate,[26] but just think about the potential of a device like this: no more bulky monitors or sensors applied with sticky and often skin-irritating gels. Besides monitoring

heart rate and pulse rate, imagine if an electrical "skin tattoo" could help you heal faster? Imagine the potential of a wearable electric bandage for any open wound that makes your skin tissue grow and regenerate without the need for stitches or surgery. Welcome to the age of electrically precise medicine.

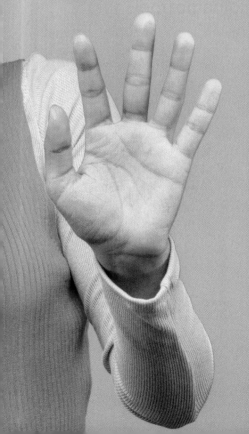

The idea is to tap into the body's natural electrical fields during healing. When there is a wound, our cells push ions through their membranes to generate an electric field—basically a signaling for the cells to line up and start growing in the direction of the wound to start the repair process. Drawing on the body's wound-healing response, materials scientist Xudong Wang and his colleagues at the University of Wisconsin–Madison developed an electric bandage that converts mechanical energy emitted by a patient's body motions into a healing charge that the body can absorb and use to jump-start its own electrical healing signals. The premise is based on triboelectricity, a reaction that causes what we know of as "static electricity." Ever get a shock when you touch a surface or a piece of clothing after, say, rubbing your hands together? Basically, the zap you feel are the surface-stealing electrons you generated when you rubbed your hands together. The bandage that Wang and his team are testing is based on this electrical stealing—in this case, from your skin to the cells that need to start repairing. So far in lab tests they have seen healing periods in rats shortened by as many as nine days.[27]

We are just at the opening gate of the new frontier of medicine using electrical frequencies to disrupt communication and fight infection. Some exciting new work has uncovered how to beat bacteria at their invasive game with an electrical bandage to stop them from creating slimy films known as biofilms over wounds. Biofilms are like an army creating a fortress or bunker, defending the bad bacteria in the wound from antibiotics and preventing the good bacteria from coming to the rescue. A team at Ohio State University's Center for Regenerative Medicine and Cell-Based Therapies created a bandage that has silver and zinc printed onto it, which help to generate a weak electric field when moistened from your body's perspiration or blood. The electricity in the bandage disrupts the bacteria signaling—known as quorum sensing—which, in the case of a wound, is how the bad bacteria communicate with each other to colonize the wound area and cre-ate that biofilm. The team is currently investigating whether these bandages can fight multidrug-resistant bacteria.[28]

Ingestible, Tunable, Electrical Medicine

The appeal of ingestible, biocompatible, and biodegradable electrical meds and wearable electrical bandages lies in their ability to bring precise, personalized treatment to patients. Unlike conventional drugs that are taken orally and end up in nearly every cell in the body before making their way to their intended target, electroceuticals are expected to hit their mark. And as health care becomes a consumer business, patients are eager for more customized approaches (think *bespoke treatments*). This will, of course, be incredibly disruptive to the health-care industry as we know it, analogous to what happened in the music industry when it went from vinyl records to cassette tapes to CDs to digital recordings and had to adjust to a world where music lives in the clouds and can be accessed on demand. Perhaps one day our daily supplements or drug dosages will live in our personal medical cloud, accessible with an app that speaks to our wearable or ingestible health device. This is one way to think of how innovations in bioengineering will drive electricity-based therapies, giving individuals control of their health care therapies and the power to charge, stimulate, record, and deliver treatments with remarkable precision.

What if you, the patient, had the opportunity to control your "dosage" of therapy—*inside* your body—through your smartphone? Need to stimulate your T cells for an immunity boost? There may be an app for that in the not too distant future. Any number of

treatments—pain management, immune therapy, or even gene therapy—could potentially be controlled remotely. Wirelessly activated medication could offer a nearly instantaneous on-off switch, in contrast to conventional drugs, which can take hours to act and which linger in the body. Researchers are currently investigating a way to use radio fields to trigger cells to supply therapeutic proteins that are costly to deliver by drugs or other means.

A team from Rockefeller University and Rensselaer Polytechnic Institute developed a new system dubbed "radiogenetics," which could make it possible to control cells and genes in living animals without wires, implants, or drugs. So far, the researchers have successfully used electromagnetic waves to turn on insulin production to lower blood sugar in diabetic mice. Their system relies on a natural iron-storage particle, ferritin, that when exposed to a radio wave or magnetic field opens up an ion channel in the body, leading to the activation of an insulin-producing gene.[29] Other techniques to remotely control the activity of cells or the expression of genes in living animals already exist but have limitations. For example, systems that use light as an on/off signal require permanent implants. The potential of remote regulation of gene expression using wireless, low-frequency radio waves (RFs) is just the beginning of a new generation of nanoscale wireless systems for medical therapeutics.

It is perhaps not too far-fetched to imagine drugs of the future as frequency nanodevices that are wirelessly programmable by a smartphone app or voice-activated ("Siri, start my insulin shot").

Or better yet, will we just need to plug in our ear pods and listen to our medicine?

The Power of Sound

For centuries, music has moved us humans and, as we established in chapter 2, the natural world has rhythms it moves to. Music is the universal language, and you have probably witnessed how a roomful of strangers can start grooving to a beat. Music also exerts a pull on our emotions—admit it, you have probably laughed at a meme of a toddler bouncing to a pop tune, or felt your heart leap with joy at an aria, or been moved to a tear or two by a tune that, well, just makes you melancholy. Music is an art form, an arrangement of sound waves and silence that makes the tones pleasing (although some people may question whether a certain tune is pleasing or not).

Sound is the first stimulus we sense. A baby usually reacts to a voice or sounds at about week twenty-five or twenty-six in the womb.[30] Hearing is also thought to be the last sense we leave this Earth with, which is why people are often encouraged to speak to their loved ones in the twilight moments before death.[31]

Australia's Aborigines were the first ancient people known to use a form of sound healing, dating back about forty thousand years. They used the didgeridoo, a wind instrument that looks like a long wooden tube made from fallen eucalyptus branches that were hollowed out. When blown through, the didgeridoo produces low, buzzing frequencies that were used for healing and ceremonial rituals. The Egyptians also understood the power of sound for healing the sick, often using seven vowels as their sacred sounds that together converged into a healing harmony. Egyptian high priestesses also played the sistrum, which looked like an ancient tambourine. The sistrum was thought to generate an ultrasound to enhance healing, and it is still in use today.[32] Pythagoras, the Greek father of mathematics, also believed in the power of sound. In fact, one of the famous quotes from his writings in the fifth century BC talks about the celestial connection of sound and the human body: "Each celestial body, in fact each and every atom, produces a particular sound on account of its movement, its rhythm or vibration. All these sounds and vibrations form a universal harmony in which each element, while having its own function and character, contributes to the whole."[33]

North American indigenous cultures, too, were known to use sounds and human vocals to evoke healing. Drums, flutes, and other percussive instruments were accompanied by a form of vocal chanting, and this is still used in energy-healing practices today. Chanting is found in many ancient and traditional cultures around the world, used like a form of prayer or affirmation, and usually for spiritual development. In ancient Chinese and Tibetan cultures, as well as Vedic and Hindu practices, chanting was a route to spiritual growth and physical well-being.[34] Chanting was used in prayers from the early days of the Roman Catholic Church (the Gregorian monks) to Qur'an readings to Jewish cantillation to Hindu mantras. Today, any armchair yogi will know the vibrational warmth of a slow-releasing "om."

Psychoacoustics: The Relationship Between Sound and Sensation

There's a science to how chanting affects our physical, mental, and emotional wellness. A recent study published in the journal *Scientific Reports* found that religious chanting can induce positive feelings and calmness, which lead to relaxation. Compared to the resting state, studies have shown that religious chanting increases the stability of cardiac activity and regulates the cardiovascular system, which illustrate the mechanisms that can be used to help with positive stress-reducing effects.[35]

For example, using multimodal electrophysiological and neuroimaging methods, the researchers at the Buddhism and Science Research Lab, Centre of Buddhist Studies at the University of Hong Kong were able to illustrate that the brain experienced a change during religious chanting, namely in the midline of the brain, where the limbic lobe sits, the region that crosses brain hemispheres and houses the amygdala and

hippocampus and plays a role in regulating emotions and memory. The chant used in their test was the Buddha Amitābha, which involves repetitively reciting the syllables comprising the name of Buddha Amitābha, either silently or aloud. They discovered that during intense chanting of Buddha Amitābha, there was an increase of delta-band power in the brain, the slow waves associated with deep sleep or meditation, usually in the frequency range between 0.5 and 4 Hz. Researchers suspect that this could be beneficial to people who suffer from sleep disorders. Similar to sleep, delta-band activity has been suggested to help with a universal response to injury.[36]

In other words, chanting can possibly make you sleep better, balance your heart rate, help your body respond to injury, and potentially reduce stress. Try a little test that you yogis will recognize: All together now, "Oooommmmmmm . . ."

Did You Know That You Can Tune Your Own Body?

Alternative-healing practitioners, namely those in sound healing, work with the idea that certain frequencies can induce a healing effect. They generally believe that all humans are tuned the same, yet each human has his or her own unique tone, which is expressed as a combination of the five elements—ether, air, fire, water, and earth. Some people are faster and some are slower by nature, and your own unique tone is what is called "your nature." Sound-healing treatments usually involve listening to percussive instruments such as gongs, Tibetan singing bowls, and tuning forks, used in group sessions such as sound baths, or in individual treatments. Sound therapies are sought out as alternative treatments for problems such as anxiety, chronic pain, sleep disorders, and PTSD.

BioSonic Repatterning™, developed by Dr. John Beaulieu, uses tuning forks to realign the nervous system, discharging emotional energies and helping to keep your energy moving in a productive way. His company, Biosonics, develops and distributes life-changing products on music and sound healing, energy medicine, and higher consciousness, including tuning forks. Beaulieu uses the analogy of tuning a piano to understand the effects of BioSonic Repatterning: all pianos are tuned the same; however, once in tune they are vehicles for expressing different types of music, and the same is true for us. When we are "in tune" we will be attracted to different types of "life music."

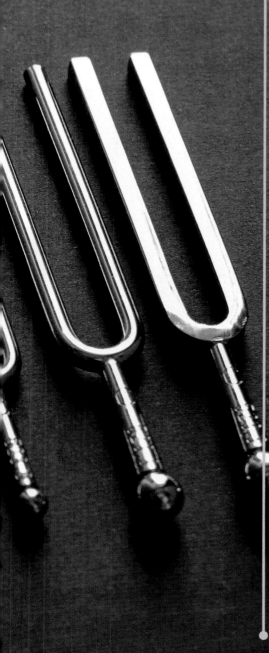

BioSonic Repatterning and polarity therapy are based on ourselves as vibrational beings. Dr. Randolph Stone, the founder of polarity therapy, used the term "ultrasonic core" to describe the vibrational core of our being. It's been called other names that include one's fundamental tone, one's essence, or your soul. Biosonic means "life sound" and it is using sound in all forms to bring one into resonance via sonic alignment with the ultrasonic core.

—John Beaulieu, ND, PhD,
sound healing pioneer,
Sputnik Futures interview, 2002

I attribute my success in the corporate world to my early studies in sound and energy medicine. Having been a student of Dr. Jeff Thompson and Dr. John Beaulieu starting over twenty-five years ago, I have learned how to stay in tune with myself, others, and the natural world. Entering the corporate world, I realized that there is real science when we say "We are on the same wavelength" and so I set out to do things such as finding out the fundamental frequency at which a group of employees resonated and fed that sound back to them while we created product together, allowing them to do their best work. We are all vibrational beings and there is unlimited potential in using sound to connect us to our original blueprint for health and well-being and to be in resonance with each other, enabling great things to be achieved.

—Ivy Ross, business executive, jewelry designer, sound practitioner, and vice president of hardware design, Google

You may know a tuning fork
primarily as an implement for
tuning musical instruments: a two-
pronged metal fork, and the thick-
ness and length of the tines has
to do with the frequency of the
note it resonates with, such as
C, G, E, or F. Tuning forks work
by releasing a perfect wave pattern, which a musician works to
match using his or her own instrument. When the instrument is out
of tune with the fork, you hear in-
terruptions in the wave pattern;
when the instrument is in tune, the
sound resonates with the tuning
fork, flowing smoothly and harmo-
niously.

This same wave pattern is leveraged in the testing and diagnostics
of physical disorders such as testing certain types of hearing loss,
called a Rinne test. The Rinne test
involves a doctor placing a humming tuning fork near the patient's
skull, and then near the patient's
ear, using a stopwatch to record
how long the patient can hear it.[38]

A tuning fork is also used as an
in-the-moment diagnostic tool
when X-rays are not readily available, to determine whether a
person—say, a runner in a race—has
a stress fracture. To assess this, you
need to strike the tuning fork so it

starts resonating, then place it as close to the stressed bone area as possible. The high-frequency vibrations from the tuning fork should travel into the bone. If the injured person experiences sharp pain, there's a possibility that a stress fracture exists. In his article in *RunnersConnect*, John Davis highlighted a 1997 peer-reviewed study that examined the efficacy of using a tuning fork to assess stress fractures. The procedure was conducted by Emil Patrick Lesho, a military doctor stationed at Fort Richardson, Alaska. Lesho used a 128 Hz tuning fork to test a tibial stress fracture, followed by a traditional bone scan. He concluded that if there was no pain felt by the patient, the tuning fork test wasn't sensitive enough to rule out a stress fracture; however, Lesho did suggest that "a positive tuning fork test might be enough to justify treating a patient for a stress fracture immediately instead of waiting for advanced imaging results,"[39] suggesting a diagnostic use of vibrations prior to the healing therapies.

In energy medicine and sound healing, therapists use calibrated metal tuning forks to apply specific vibrations to different parts of the body, meant to help release tension and promote emotional balance. It works similarly to acupuncture, using sound frequencies for point stimulation instead of needles. But in 2002, Beaulieu and Dr. George Stefano conducted research to show the biological healing effects of using biosonic tuning forks. Using a petri dish, they demonstrated the ability of specific vibrations (using the Biosonic Otto 128 cps tuning fork) to spike nitric oxide levels. Nitric oxide is a gas our cells release to encourage the cardiovascular system to relax, which could help prevent heart attacks, and also works with our immune system to destroy pathogens. The in vitro findings led them to theorize that a specific vibration, when combined with a chemical response (possibly from the brain), could move throughout the body, potentially creating a relaxation response. The important benefits of the relaxation response in preventive medicine, yoga, medication, mindfulness, etc., have been well documented in many research articles.[40]

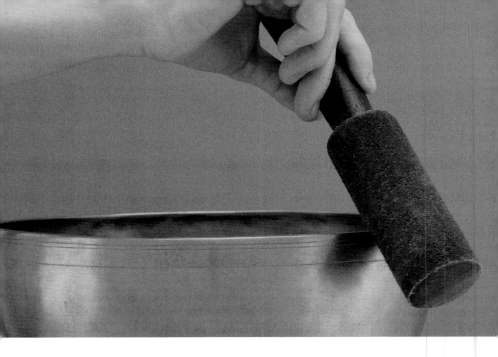

I discovered BioSonic Repatterning while sitting in an anechoic chamber at New York University. The chamber is a completely soundproof room that resembles a sensory deprivation chamber. I had read about the experiences of the composer-philosopher John Cage and decided to conduct a similar experiment. While in the chamber, Cage heard two sounds, one high-pitched and the other low-pitched. The engineer he was working with informed him that the high sound was his nervous system and the low sound was his blood circulating. Inspired by John Cage's experience, I sat in an anechoic chamber for five hundred hours over a period of two years and listened to the sounds of my own body. I began to correlate different states of consciousness with the different sounds of my nervous system. Being a trained musician, I noticed that the high-pitched sounds of my nervous system consisted of several sounds in different intervals. Then one day I brought two tuning forks and tapped them. Immediately I observed that the sound of my nervous system came into resonance with the sound of the tuning forks. It was then I realized that people can be tuned like musical instruments!

—John Beaulieu, ND, PhD, sound healing pioneer, founder of Biosonics (via biosonics.com)[41]

Other forms of sound healing include sound baths, group experiences in which participants lie on the ground, usually in a circle (or concentric circles), as therapists/instrumentalists play soothing tones that "wash" over the group to help relieve tension. The instruments used range from gongs to Tibetan singing bowls that immerse participants in an array of sound waves said to create a calming and healing effect that is induced by the frequency the instruments generate.

The underlying science relies on the fact that brainwaves can be modified by externally produced sound frequencies in a process called entrainment, where one frequency is synchronized with another frequency.[42] Entrainment implies that sound can be used to tune brainwaves to specific frequencies to help reach a desired state of mind.[43] Theoretically, the synchronized sounds of singing bowls or gongs envelop you in one harmonious and continuous frequency that may help adjust or calm your thought patterns.

The science validating the biological effects of tuning forks and other means of sound therapy is unfortunately still sparse, yet there are organizations such as the Consciousness and Healing Initiative that are creating means for scientists, health care practitioners, educators, and innovators to further the understanding and real-world applications of subtle-energy healing practices, including sound.

If you have experienced a sound bath or a gong session, you can't deny the human condition—that centered, balanced, and calm state felt during a sound-therapy experience. In this age of always being on, sound is having a resurgence for the self-care set to reduce stress, social and ecoanxiety, sleeplessness, and more.[44]

Chanting versus Listening

In a small 2012 study, Helen Lavretsky, a professor of psychiatry at UCLA, and her team looked at the effects of listening to music versus chanting to help alleviate depression, and found that chanting improved the outcome. The researchers split thirty-nine people caring for family members with dementia into two groups. One group was tasked with listening to relaxing music for twelve minutes each day for eight weeks. The other group was asked to practice Kirtan Kriya, a meditative form of yoga that involves chanting, for the required twelve minutes each day for eight weeks. The chanting group reported feeling better, with 65.2 percent describing fewer depressive symptoms, whereas the group that listened to relaxing music reported about a 31.2 percent improvement in their depressive symptoms. Both felt better, but the chanting did raise the bar.

—Quartz[45]

Integratron Sound Baths

The Integratron, built in about 1954, is a one-of-a-kind thirty-eight-foot-high, fifty-five-foot-diameter, all-wood dome that was designed to be "an electrostatic generator for the purpose of rejuvenation and time travel," according to its website. Located twenty miles north of Joshua Tree National Park in Landers, California, Integratron is said to be built on "an intersection of powerful geomagnetic forces that, when focused by the unique geometry of the building, amplify the Earth's magnetic field." The website claims that magnetometers can read a significant spike in Earth's magnetic field in the center of the Integratron. But to believe this, you have to experience it, and a team from Alice in Futureland did, participating in a sound bath that included thirty minutes of twenty quartz crystal singing bowls played live in the center of the dome-shaped structure.

—Integratron[46]

Will Music Be a Drug?

While music therapy is often lumped in with sound therapy, there's a distinction in that the evidence backing music therapy is more robust. Studies show that humans are hardwired for music[47] and that music activates many global regions of the brain. Researchers are now using biofeedback and machine learning to identify how music's structural properties (such as beat, key, and timbre) specifically impact biometrics such as heart rate, brainwaves, and sleep patterns, developing a new "wellness music" as precision medicine.[48]

There are some promising studies showing how music can help heal, encourage social interactions, trigger memory, lower depression and anxiety, work alongside anesthesia during spinal surgery, and even help premature babies gain weight more quickly.[49]

The music-therapy field officially emerged in 1999 as neurologic music therapy (NMT), attracting start-ups who leverage digital technology with the new tools of tracking our brainwaves. NMT is based on music's ability to stimulate a whole range of different parts of the brain. "There's no other stimulus on earth that provides such a global activation of our brains as music," explains Brian Harris, cofounder of MedRhythms, in the article "The Doctor Is in (Your Pocket): How Apps Are Harnessing Music's Healing Powers" on Pitchfork.com. Another principle of NMT is that music helps with neuroplasticity, the ability of the brain to change over time with training. Science used to think that at a certain age the brain stopped developing and changing, but neuroscientific research has found that the brain can be trained—hence the term "neuroplasticity." A data-driven review by Northwestern Univer-

sity researchers in 2010 pulled together converging research linking musical training to learning, showing effects on skills such as language, speech, memory, attention, and vocal emotion.[50] MedRhythms is a start-up in Boston developing treatments based on research into the effect music has on nonmusical parts of the brain, such as language, cognition, memory, and movement. The company's team of therapists works with people recovering from brain injuries, diseases such as Parkinson's and Huntington's, and stroke. Their goal is to create an entire digital medicine platform.[51]

Perhaps the most known advocate of music therapy for medical intervention is the late Dr. Oliver Sacks, former professor of neurology at Columbia University and author of the book *Musicophilia*. His account of using music therapy for Parkinson's disease is what inspired the book and 1990 film *Awakenings*.[52]

Several studies have documented how music can aid speech recovery. Dr. Teppo Särkämö of the University of Helsinki found in 2008 that when stroke patients were given musical recordings, their linguistic capacities improved to a greater degree than patients who were given audiobooks. By examining MRI scans of stroke patients, he has shown how music also induced visible changes in some brain structures after just six months of treatment.[53] Another example of the power of songs to help with language recovery is that of US congresswoman Gabrielle Giffords, who was shot in the head in 2011 but survived. Giffords suffered from aphasia–the inability to speak because of damage to the language pathways in her brain's left hemisphere. She worked with Meaghan Morrow, a music therapist and certified brain-injury specialist at TIRR Memorial Hermann rehabilitation hospital in Houston, and after ten months was able to regain her speech, training her brain to layer words on top of melody and rhythm.[54] Music aided language recovery better than just practicing forming words itself.

Music therapy has even had some success helping to stimulate motor skills. Dr. Simone Dalla Bella, from the Department of Psychology at the Université de Montpellier in Montréal, is study-

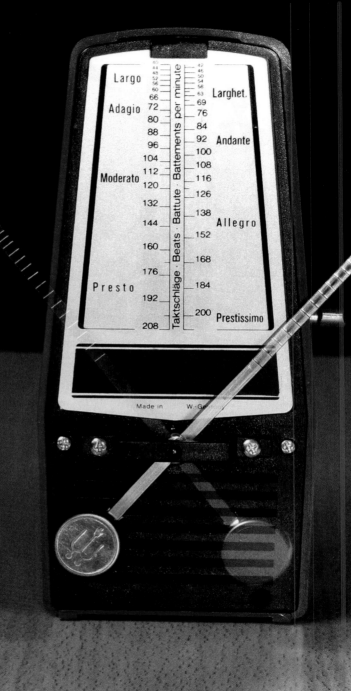

ing what he calls "melodic gait therapy," to help people with Parkinson's disease walk more steadily. The approach uses metronomes (a device used by musicians that marks time at a selected rate by giving a regular tick) and rhythms of percussive instruments. About twenty-one patients were put through three weeks of training, the music stimulation adjusted to each patient's preferred cadence (gait frequency). The study took place at the Clinic of Cognitive Neurology at the University Hospital of Leipzig, Germany, where patients were asked to walk along with a familiar German folk song ("Hoch auf dem gelben Wagen") played without lyrics. The results were promising: patients increased their gait speed and stride length significantly after the training, and the effects were still significant one month after training. The researchers also found that some patients had improved motor control in addition to a better gait.[55]

You've Got the Music in You

We talked earlier about how chanting and singing can help relieve stress and anxiety, which is steadily becoming a serious mental health issue worldwide. The stress and overload of our always-on work lifestyle, what we call "burnout," is officially a medical condition, according to the World Health Organization.[56] People seek out music and sound therapy, whether through sessions or the many mindfulness apps that feature a mantra or sound meditation to help alleviate burnout and the stress that causes it. But there is a segment of the population that has just as serious mental health issues as those still in the corporate or gig-economy race, and that is the elderly: there were 703 million people age sixty-five or older in the world in 2019.[57] Elderly individuals who have chronic health disorders have a greater chance of having symptoms of depression, also a serious mental health issue: 7.7 percent of adults age fifty or older in the United States reported currently being depressed, and 15.7 percent reported chronic depression.[58] According to the National Council on Aging, 7 million older Americans are affected by depression and many do not receive the treatment they need. Add to this the fact that about 92 percent of older adults have at least one chronic health disorder, such as heart disease, cancer, stroke, or diabetes, and 77 percent of older adults have at least two.[59] According to a study published in the peer-reviewed *Nursing 2020*, chronic illnesses are what compromise the quality of life (QOL) for older adults, both physically and emotionally.

Western medicine turns to pharmaceuticals to alleviate the symptoms and struggles of depression, but the medications used to treat the symptoms of depression are often costly and may have adverse effects when mixed with the other critical medications a person may be taking. Jenny Quach, BSN, RN, and Jung-ah Lee, PhD, RN, did a systemic review of the various literature available that has shown that music therapy can improve mood and behavior in older adults with dementia. They found that eight of nine studies that specifically used a depression-measuring instrument showed significant decreases in depression. All studies reviewed showed some benefits of music therapy in improving emotional well-being in older adults with chronic diseases. Listening to music, playing an instrument, singing, or a combination of these were useful in relieving depression and improving overall mood.[60] They concluded that nurses and health care providers should be aware of the benefits of music therapies and consider incorporating them into patient care when feasible.[61] Music therapy is often a low-cost alternative to expensive medications. And don't we all, no matter what age, love a little music?

Classical Music Can Destress You

A study conducted by a team from the department of clinical psychology and psychotherapy at the University of Zürich in Switzerland found that sixty healthy female subjects with a mean age of 25.3 years who listened to classical music before a stressful event recovered from the stress faster than those who listened to rippling water or simply relaxed quietly.
—*PLoS One*[62]

Take Your Music Daily

Researchers examined the potential stress-reducing effect of music listening in everyday life using both subjective and objective indicators of stress. Fifty-five healthy university students participated in the study, during a regular term week (five days) and during an examination week (five days). The participants rated their current music-listening behavior and perceived stress levels four times per day, and a subgroup also provided saliva samples to analyze the levels of cortisol. The results revealed that daily music listening was effective in reducing subjective stress levels.

—National Institutes of Health[63]

Why You Get Chills From a Certain Part of a Song

There's a name for the sudden burst of cold that you may experience during one of your favorite songs—frisson—and it stems from the nerve fibers that connect the brain's auditory cortex (the part of the brain that processes sound) to the anterior insular cortex (the part of the brain that processes emotion). If you experience frisson, as about half to two-thirds of the population does, it means that the connection between those two cortexes is strong. When we listen to music, our brains continually process melodies and predict the recurrence of a repetitive musical phrase. When something unexpected (but pleasing) happens in a given song, these cortexes react. For some, this response may lead to a physical, emotional sensation: chills

Why You Can't Get That Song Out of Your Head

It all has to do with something called the exposure effect. This hypothesis holds that your brain experiences positive psychological effects when it encounters something it already knows, such as a repeated melody, beat, or chorus. The effect is so powerful that, in terms of activating our brains' reward centers, repetition even trumps our personal musical preferences. And sometimes, for that reason, a repetitive song—and one that you don't even like—will stick around in your brain longer than you'd want. This phenomenon is known as an earworm.

—Erin Kelly, *All That's Interesting: The Science Behind Music*[64]

> One good thing about music, when it hits you, you feel no pain.

—Bob Marley, musician[65]

Music as Precision Medicine

Let's face it: we all have self-medicated on a playlist at one point in our life (hello Breakup Mixtape!), but music actually offers one of the most promising approaches to pain management. One region of the brain that music stimulates, called the *nucleus accumbens*, is the same region that releases dopamine, the feel-good neurotransmitter associated with pleasure. This is the same area that neuroscientists say may be responsible for the "chills" we get when we listen to a particularly pleasing piece of music. First our brain processes the tune through the caudate nucleus, which senses and anticipates the buildup of our favorite part of a song, getting our brains fired up. Then the nucleus accumbens is triggered, causing the release of endorphins—neurotransmitters that interact with the opiate receptors in the brain to reduce our perception of pain, basically acting like a drug.[66]

Sync Project, one of the leading personalized music platforms, is collaborating with top neuroscientists and musicians on large-scale studies that also includes citizen musicologists to measure how the structural properties of music—such as beat, key, and

timbre—impact our biometrics of heart rate, brain activity, and sleep patterns. Applying machine learning to this data, Sync Project (recently acquired by the speaker maker Bose Corporation) is out to prove the healing effects of personalized musical therapeutics for everything from pain management to improving sleep.

One of Sync Project's studies is looking at music's effect for ameliorating pain, and its potential as an adjunctive analgesic. This is in line with recent research showing that music affects the same neural pathways that are regulated by psychostimulants and other drugs. And this couldn't have come at a better time, with the increasing epidemic in opioid use in the United States and globally. What is drastically needed is an affective adjunctive therapy to opioid-based analgesia that is a behavioral intervention to opioid-based pain treatment. Music is one such adjunctive therapy that actually resonates with the opioid-using population. Music helps modulate the perception and experience of pain that your brain responds to by blocking the messages of "pain" to help patients who are addicted to opioids and other pain meds transition to a lower dose of their pharmaceuticals.[67]

Recovery Unplugged

One addiction care organization that seeks to make "rocking out" a cutting-edge tool in the treatment of substance abuse is Recovery Unplugged. They combine traditional elements of treatment such as detox and counseling with musical therapy techniques to help "people reclaim their lives from drugs and alcohol." Recovery Unplugged uses music-assisted treatment as a nontoxic, noninvasive therapeutic option during addiction treatment, and has experts on staff who can apply these proven music interventions to replace the feelings one used to get from alcohol and drug use.

—Recovery Unplugged[68]

The World's Most Relaxing Song

"Weightless" is a song written in 2012 by the UK band Marconi Union to specifically reduce anxiety, blood pressure, and heart rate. It was composed of dreamy, mellow synths, soothing piano and guitar melodies, and electronic samples of natural soundscapes. And it was put to the test as a preoperation musical drug. In a study of 157 patients, the song performed well as a sedative given for three minutes while having an anesthetic to numb the part of your body being worked on. Research from Penn Medicine concluded that "music is a viable alternative to sedative medications in reducing patient anxiety prior to an anesthesia procedure."

—*Penn Medicine News*[69]

Take a Sonic Mushroom Break!

Everyone needs an audio break, time to tune in and turn off.

We at Alice in Futureland hear you. Partnering with leading composers and sound designers, we created Sonic Mushrooms—Neuroaudio to help increase focus, enhance thinking, or just chill out. You can visit aliceinfutureland.com to sample one.

And the Beat Goes Binaural

The deep reach of YouTube and online music stations such as Spotify are introducing us to a new kind of alternative music therapy: binaural beats. What makes this sonic experience unique is that the right and left ears each receive a slightly different frequency tone, yet the brain perceives these as a single tone. Trippy, eh? Listening to binaural beats for a recommended period can affect a person's behavior and sleep cycle. And there's a science to it (of course).

Binaural beats were first discovered by the meteorologist Heinrich Wilhelm Dove in 1839. Among Dove's accomplishments—which include writing more than three hundred research-based papers on everything from magnetism to distribution of heat over the Earth that profoundly influenced our modern study of climate change—he is perhaps most cited for a paper he published about a study he did while teaching in secondary school. Dove put a student in a room, and on one side, he placed a tuning fork, with a listening tube that ran from the fork to the student's ear. He placed an additional tuning fork on the other side of the room, again running a tube up, this time to the student's opposite ear. Dove reported that even though the two forks didn't vibrate at the same frequency, the student could hear the difference but as one combined sound, a slow beat that is now known as a binaural beat.[70]

It wasn't until 134 years after Dove's paper that Gerald Oster's article "Auditory Beats in the Brain" (*Scientific American*, 1973) presented a series of relevant research and laboratory findings on binaural beats, and the scientific community began to pay attention. Oster described how the rhythm of the binaural beat equals the difference between the two tones and, if sustained, that this rhythm can be entrained throughout the brain. This was an important discovery, because it showed that if the right frequency is selected, it can produce particular states of electrical activity in the brain.[71]

A pilot study in 2001 showed the success of binaural beats inducing brainwave states to help decrease anxiety in patients suffering from chronic anxiety. Binaural beat audio rich in delta brainwave entrainment can be an intervention that inhibits anxiety.[72]

In a study published in *Anaesthesia* in July 2005, Dr. David Laws and his team presented a case where binaural beat audio was used successfully to help with preoperative anxiety in patients undergoing general anesthesia for day-case surgery.[73] Studies have demonstrated that preoperative anxiety may increase the anesthetic requirements during surgery[74] and may increase postoperative pain.[75]

Brainwave Entrainment Is Like Going to a Gym for Your Brain

There are several studies demonstrating the sonic brain power of binaural beats, but what's interesting is the explosion of popularity it is having for people taking brain entrainment into their own hands. A simple way to understand brain entrainment is to think of CrossFit: you use a range of different tools such as weights or rubber tires to train different muscle groups. Brain entrainment is the process of training your brainwaves by using pulsating sound—in this case, binaural beats. Spotify has 250,000 (and counting) monthly subscribers to its Binaural Beats playlist. On YouTube, you can choose from a mood menu of binaural beats for deep focus, super intelligence, rapid healing, creativity, REM sleep, to cleanse infections, to attract abundance—you get the picture.

Medical News Today lists the frequency and benefits of binaural beats on the brain:

DELTA PATTERN: Binaural beats in the delta pattern operate at a frequency of 0.5–4 Hz with links to a dreamless sleep.

THETA PATTERN: Practitioners set binaural beats in the theta pattern to a frequency of 4–7 Hz. Theta patterns contribute to improved meditation, creativity, and sleep in the REM phase.

ALPHA PATTERN: Binaural beats in the alpha pattern are at a frequency of 7–13 Hz and may encourage relaxation.

BETA PATTERN: Binaural beats in the beta pattern are at a frequency of 13–30 Hz. This frequency range may help promote concentration and alertness. However, it also can increase anxiety at the higher end of the range.

GAMMA PATTERN: This frequency pattern accounts for a range of 30–50 Hz, promoting maintenance of arousal while a person is awake.

Give Your Brain a Workout

One of the leading researchers in brainwave entrainment is Dr. Jeffrey Thompson, founder/director of the Center for Neuroacoustic Research (CNR) in Carlsbad, California. He has worked with NASA and other groups on gentle sound modalities for managing health and wellness, and for personal growth and transformation. He has developed an individualized sound therapy called Bio-Tuning/Sonic Induction Therapy, which builds on the phenomenon of brainwave entrainment. In his recordings, there are subliminal frequency patterns woven in every sound. When your brain senses these hidden pulses, your brainwaves will change to match them, enhancing your ability to enter into a meditative state. Listening repeatedly to a personalized therapeutic sound track can help enhance brainwave function—like going to a gym for your brain![76]

We still have so much to learn about the brain and consciousness. What we think of as our or-

Benefits of Binaural Beats

The purpose of using binaural beats therapy may differ among individuals. Some people may need help decreasing their anxiety, while others might want to improve their concentration or deepen their level of meditation. Proponents of binaural beat therapy suggest that the potential benefits include:

· reduced stress and anxiety;

· increased focus, concentration, and motivation;

· improved confidence;

· better long-term memory after exposure to beta pattern frequencies;

· deeper meditation; and

· enhanced psychomotor performance and mood.
—*Medical News Today*, September 29, 2019[77]

For the Record

A 2019 review of twenty-two studies found a significant link between more prolonged exposure to binaural beat tapes and reduced anxiety. It also found that practitioners did not need to mask the beats with white noise for the treatment to have an effect.
—National Institutes of Health[78]

Power of Ten

Some researchers have suggested that ten minutes of exposure to a 6 Hz frequency could induce a brain state similar to that during meditation.

—*Frontiers in Neuroscience*[79]

dinary consciousness actually represents only about 5 percent of our brainpower. The remaining 95 percent represents the default brain function, what Thompson calls the default mode network (DMN). Emerging research is looking at this brain network as the physical place where the mind, emotions, and body come together—potentially where the "mind-body" resides.

When our brainwaves are not synchronized, this can lead to debilitating conditions such as schizophrenia, anxiety syndromes, depression, PTSD, Parkinson's disease, Alzheimer's disease, autism, classic psychiatric dissociative disorders[80]—many of the conditions we have been covering in this chapter that are responding to music therapy.

Cosmic Tunes

Now you can entrain your brain with the meditation frequency of the universe. Thompson has taken recordings directly from the NASA *Voyager I* and *Voyager II* space probes as they passed near different planetary and lunar bodies within our solar system and compressed them electronically down to their fundamental harmonic frequencies. These sounds are from the interaction of the solar wind and the ionosphere of each of the outer planets in Earth's solar system. For example, he has soundtracks of Saturn's rings, the rings of Uranus, and the song of Earth.[81]

Tickle Your ASMR

Even though you may not recognize it, you would have to be living totally offline not to have heard ASMR in a commercial, YouTube video, or even a Billie Eilish song (in fact, she may just be the reigning queen of ASMR music!). It can make your skin crawl, the back of your neck tingle. ASMR, autonomous sensory meridian response, is commonly described as having a narrow range of specific sensations and feelings. Sensations include tingles, chills, and/or waves in the head and possibly other areas. Feelings include euphoria, happiness, comfort, calmness, peacefulness, relaxation, restfulness, and/or sleepiness. The stimuli or "triggers" for ASMR mostly fall into three main categories: tactile, or touch (think of light tickling); visual, such as observing very slow hand movements; or auditory stimuli, basically sounds that are whispery soft, or the sounds of eating, brushing hair, scratching, or crinkling. ASMR sounds are the rage these days; just check out the hashtag #ASMR on Instagram and you'll see 6.2 million posts. Some ASMRists (ASMR art-

Binaural Beats + ASMR = Sleep

A study published in December 2019 proposed a novel way to use sounds to induce sleep. The researchers leveraged the power of binaural beats that induce brains signals at a specific desired frequency but realized that the stimulus may be a bit uncomfortable for users to listen to while trying to go to sleep. Their idea was to exploit the feelings of calmness and relaxation that are induced by ASMR, using ASMR triggers of natural sounds. The combined auditory stimuli (the 6 Hz binaural beat and ASMR triggers) did induce the brain signals required for sleep, while simultaneously keeping the user in a psychologically comfortable state.[82] This means that it might be time to trade in your nature-sound machines for ASMR binaural beats.

—*Frontiers in Human Neuroscience*

ists) have upward of 5 million followers on YouTube.

Many believe that ASMR can be used as part of therapies for mental health or stress management. Data from a 2015 study by Emma Barratt and Nick Davis showed that among the participants, 98 percent who watched ASMR videos claimed to feel relaxed, 82 percent said they helped them sleep, and 70 percent used them to reduce stress.[83] The same study found an increase in "flow," or concentration during a task, when listening to ASMR. Data obtained also illustrates temporary improvements in symptoms of depression and chronic pain in those who engage in ASMR. The Barratt and Davis study helps to support the hope that ASMR could someday be approved as a medical treatment to help people with insomnia, anxiety, depression, or chronic pain.

The Colors of Noise

Just like frequencies of light have color, noise has its own set of sonic hues. You probably know the term "white noise"—it's that static noise, or "shhh" sound, when you are between radio channels (well, before the digital age of direct tuning) that basically cancels out all other background noise. Like white light, which contains all the frequencies in the visible spectrum, white noise contains all the audible frequencies that the human ear can hear, about 20 Hz to 20 kHz. White noise is an equal distribution of energy across all the different frequencies, which then make up a steady, low hissing sound that allows you to block out or mask the sounds that keep you up at night or distract you.[84] Sound machines that use white noise claim to help with sleeping difficulties and sleep disorders such as insomnia.

One of the newcomers to the sleep-inducing-sounds scene is pink noise. Similar to white noise, pink noise sounds a notch deeper and lower in its "shhh" tone. While there's more research on the lulling potential of white noise, a few studies are showing that pink noise can reduce brainwaves, which increases stable sleep.[85] One 2017 study published in *Frontiers in Human Neuroscience* found a positive link between pink noise as acoustic enhancement for deep sleep, which supports memory improvement.[86] So are you thinking what I'm thinking? Time to turn the pink on!

The Sound of Pink

Pink noise is more intense at lower frequencies, which creates a deep sound. Nature is full of pink noise, including wind, steady rain, and the sound of our heartbeat. To the human ear, pink noise sounds "flat" or "even."

—HealthLine[87]

Brown Noise/ Red Noise

Brown noise—named for the kind of noise produced by Brownian motion, the random movement of small particles suspended in a liquid, not the color brown itself—is also known as soft noise, or red noise. The motion occurs as a result of the impact of molecules on the surrounding medium, similar to what a rain shower does to a puddle on the ground, moving about all the small particles inside the puddle. Brown noise has more energy at lower frequencies than white or pink noise and has a damped, soft quality sound, like a waterfall or ocean waves. There is no research supporting the use of brown noise for sleep or relaxation, but even so there are millions of views of brown noise for sleep on YouTube.

—HealthLine[88] and the *Atlantic*[89]

Ultrasound: The Vibrational Healing Wave

Another use of sound waves in healing is through therapeutic ultrasound machines that generate sound waves higher than the frequencies we can hear. You may typically think of ultrasounds as an imaging test that lets you hear and see your baby in the womb, or for other internal pictures of the bladder or woman's uterus. But ultrasound can be part of therapeutic modality as well. Ultrasound therapy is a form of treatment used by physical therapists to relieve pain and promote tissue healing. The device looks something like a microphone with a transducer head that is put next to the area of pain. The therapy can be thermal or mechanical, with the difference being the rate at which the sound waves penetrate the tissues.[90]

Thermal ultrasound therapy uses a more continuous transmission of sound waves that causes healing in soft tissues by increasing the heat and metabolism in the tissue cells. Mechanical ultrasound therapy uses pulses of sound waves to penetrate tissues, causing the expansion and contraction in the tiny gas bubbles of the soft tissues, reducing tissue swelling and decreasing pain. Some studies show that ultrasound therapy can control certain types of chronic pain.[91]

Are We Frozen Music?

From neurostimulation to music as medicine to therapeutic ultrasonic waves, we are just beginning to understand the healing power of frequency. Frequency, while still experimental from a clinical perspective, remains one of the most sustainable, noninvasive, and potentially economical modalities in treating, or alleviating, a host of ailments, pains, mood swings, and disorders. And if you think about the idea that we are made of energy, that there are electrical currents running our biological systems, and that sound can create physical forms, it seems likely that frequency can be a positive force for repair, reconstruction, and renewal. Ultimately the range and intensity of these frequencies determine the effects they can have on our body and our mind. Just like a musical instrument, we too can be tuned. Johann Wolfgang von Goethe, a German poet and scientist in the late 1700s, once said, "Music is liquid architecture; architecture is frozen music." Music is frequency; we and nature resonate to its melodies. Perhaps all life, from the architecture of our bodies to the design of every leaf on a tree, is just frozen music, waiting to be turned on.

Free Energy

And the Quest for the Zero Point

Energy the Beautiful

Energy is everywhere, and so far we have been exploring its origins, its power, and its potential to affect us physically, mentally, and spiritually. Energy connects us to the cosmos, to nature, to each other, and to bacterial networks. Over the centuries, we humans have learned to harness energy to fuel our cities, our transportation, our homes, our things, and more—and often at a cost to the natural environment that sustains us. Wealth and energy go hand in hand, and a future world of greater material comforts is going to be one that uses more energy. Alternative sources of energy represent a tremendous opportunity.

Dematerialization is making the cost of many things trend toward zero—including, possibly, energy. The aim of today's energy mavericks is to use fewer natural resources and create more renewable power generation from solar frequencies, the kinetics of wind and water, and nuclear energy. Frontier scientists are setting their sights on harnessing the unlimited and abundant flow of free energy in the universal electromagnetic fields. We need to avoid making this new energy race a "Tesla moment," when Nikola Tesla's wireless power-transmitting tower, decades ahead of its time, ran out of funding. We need big thinkers developing sources of energy that can be scalable; and we will need new ethical policies to handle its abundance fairly. There is no reason to believe that we can't invent a democratic alternative to alternatives.

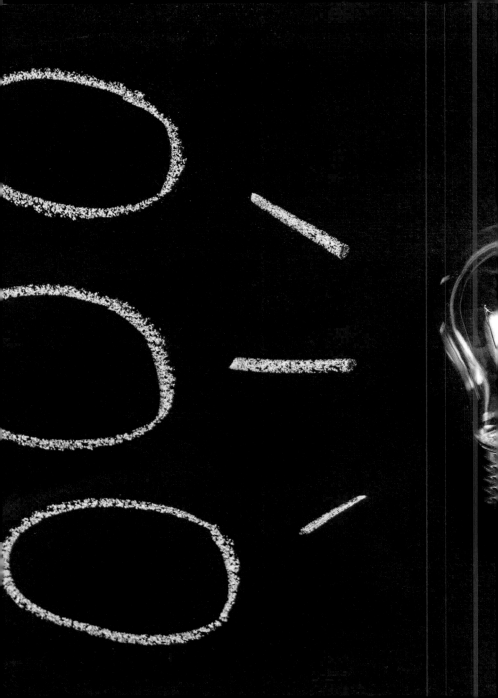

Clean Energy

For decades, forward-thinking energy experts have been exploring how we could power the planet on renewables alone. The dream of climate activists the world over is to abandon fossil fuels and adopt 100 percent clean-energy technologies. Clean energy has been around for a long time, but it used to be completely noncompetitive. Now, with the decline in the worldwide demand for coal, cleaner energy initiatives are driven by market forces rather than by regulation or government subsidies, and this trend is key to their long-term success. To halt the idea of runaway climate change, we need to squelch carbon emissions down to near zero by mid-century. That means all ideas have to be on the table, however "out there" they may seem.

Welcome to the energy frontier. Call them the mavericks, the inventors, the dreamers of dreams—Tesla was one—these are the shoulders we stand on today in a quest for cleaner, renewable, and abundant energy. A number of thought leaders are calling for a radical shift from the way research was done traditionally in the quest for new energy solutions. Dr. Gerald Pollack, professor of bioengineering at the University of Washington, is one of them. He has proposed an Institute of Venture Science that would fund large studies into revolutionary phenomena that, if any proved fruitful, would result in dramatic changes to scientific theories and applications—i.e., the world as we know it.

As far as revolutionary ideas go, aether (or ether) is at the top of the list. The word "aether" comes from the Greek *aithêr* (upper air), and according to Greek mythology, aether was the pure air that the gods breathed in the heavens. Aristotle posited that aether was the invisible, unchangeable material that permeated all the empty space in the universe. To Aristotle, aether comprised the region above the terrestrial sphere—that is, above the four elements of earth, air, fire, and water, which were subject to decay and change.

Ether is also the first of the five great elements (*Pancha Mahabhutas*) in the Ayurveda. Called

Akasha in Sanskrit, ether comes first because it is the most subtle of the elements, the essence of emptiness. Often referred to as "space," ether is the space the other elements fill.

Fast forward a few centuries, and scientists, including French metaphysician René Descartes and Sir Isaac Newton (one of the most influential scientists of all time) in the seventeenth century, to the early twentieth-century inventor Nikola Tesla (yes, the genius we have been remarking on!), looked at ether as the means to explain fundamental natural phenomena such as gravity and light. Early in his career, Newton questioned the mechanics of gravity: if celestial particles were pushing toward Earth, what was pushing these particles? But in his 1687 scientific opus *Principia*, Newton didn't discuss ether as the mechanics of gravity, putting forth instead his famous theory of the forces of attraction and repulsion.[1]

Some researchers say that ether, the invisible substance that surrounds us, could unify physics—and help us tap into an immense, infinite source of energy, one that has been called "zero point energy." (More on that in the next section.) Meanwhile, as the world awaits the end of fossil fuels and the beginning of clean, nonpolluting technologies that will serve to rejuvenate our economies, our planet, and our well-being, the call is to remove all blindfolds and consider that there may be an alternative to alternatives.

We may learn something new if we study the limitless supply of energy everywhere in the universe. If proven true, and we succeed as Tesla predicted in attaching our machinery to the ether, the wheelwork of nature, our society awaits the most intense period of evolution in the history of civilization.

Changing the
Energy Game

Our first stop on the free-energy tour is with the Advanced Research Projects Agency-Energy (ARPA-E). This is a well-funded military think tank, inspired by DARPA, that has helped invent everything from GPS to the internet. Its job is to fund the kind of transformative technologies that could change the way we view and use energy. ARPA-E funds game-changing energy technologies that are typically too early for private-sector investment—ideas from university Skunk Works and under-the-radar start-ups that are far from turning a profit today but that could pay off enormously in the future. They call themselves preventure funders who bankroll ideas that are too risky for venture capitalists.[2]

Some of the preventure ideas being funded are:

· electrofuels that use custom-designed microbes to convert carbon dioxide into liquid biofuels;

· SunShot Initiative—technologies that could lower the cost of solar power to five cents a kilowatt-hour, making fossil fuel obsolete;

· an electrical grid that could seamlessly store the power generated by the Sun or the wind—enabling renewable power to meet 24/7 demand; and

· ALPHA—fusion energy that holds the promise of cheap, clean power production. Until now, scientists have been unable to successfully harness fusion as a power source due to the high cost of research, but ARPA-E's ALPHA program seeks to create tools to aid in the development of new, lower-cost pathways to fusion power and to enable more rapid progress in fusion research and development.

—ARPA-E.energy.gov[3]

Fusion: The Holy Grail of Clean Energy?

The Sun, we know, is an abundant and free source of energy. But about a hundred years ago, scientists were just learning about the physics of sunshine, which led to what we now call nuclear fusion, or just "fusion" for short. Fusion is actually the process that powers the Sun and the other stars. It occurs when the nuclei of two atoms are forced so close to one another that they combine into one. In this dynamic process, energy is released. Scientists have been working on how to tame and safely re-create this process in the laboratory, hoping to capture what journalist Matthew Hole called "the potential to deliver near-limitless baseload electricity with virtually zero carbon emissions." Scientists have been able to trigger a fusion reaction; however, as yet it takes more energy to sustain the reaction than the amount of energy the reaction itself can produce.[4] Think of the energy from the initial reaction as just a spark, but we need a forest fire to sustain it.

That's where ITER steps in. ITER is an international nuclear fusion research and engineering megaproject being built in the South of France and will be the world's largest magnetic confinement plasma physics experiment. Its aim is to harness the power of fusion as a "potential source of safe, non-carbon-emitting, and virtually limitless energy." Thirty-five nations are collaborating to build the world's largest tokamak, a magnetic-fusion device designed to prove the feasibility of harnessing fusion energy.[5] If ITER succeeds, we will have a carbon-free source of energy based on the same energetic reaction that powers our Sun and the other stars—and which could yield cleaner, more abundant energy for all. A win-win solution that is healthier for us and less stressful for the planet.

$$\left(mc^2\alpha_0 + c\sum_i \alpha_i p_i\right)\psi(x,t) = i\hbar\frac{\partial\psi}{\partial t}(x,t)$$

$$\int_M K\,dA + \int_{2m} k_g\,ds = 2\pi\chi(M)$$

$$\frac{1}{\zeta(s)} = \sum_{n=1}^{\infty}\frac{\mu(n)}{n^s}$$

$$E_k = \frac{mv^2}{2}$$

$$G_{\mu\nu} = 8\pi G(T_{\mu\nu} + \rho_\Lambda g_{\mu\nu})$$

$$\lambda = \frac{v_{ph}}{v}$$

$$\varphi F + \frac{v^2}{2} + \int\frac{dp}{\rho} = C(t)$$

$$g = C\frac{M_3}{R_1}$$

$$E = mc^2$$

$$F = C_u v^2$$

$$E = \hbar m$$

$$E_{AT} = \psi(\Delta T)$$

$$E_k = \frac{mv^2}{2}$$

$$\nabla\cdot D = 4\pi\rho$$

$$p = m_3\frac{dr}{dz}$$

$$m\frac{d^2r}{dt^2} = F$$

$$dx_1 \dots dx_n$$

$$\varepsilon = \oint_L \vec{E}\cdot d\vec{l}\,(2)$$

$$\varphi F + \frac{v^2}{2} + \int\frac{dp}{\rho} = C(t)$$

$$i\hbar\frac{\partial}{\partial t}\Psi(\vec{r},t) = \left[-\frac{\hbar^2}{2m}\nabla^2 + V(r,t)\right]\Psi(r,t)$$

$$p = m_3\frac{dr}{dz}$$

$$\int_a^b f'(x)\,dx = f(b) - f(a)$$

Next Stop: Cold Fusion

Re-creating fusion has been a challenge because the protons in nuclei inside the atoms that need to fuse to create that energy are all positively charged. To initiate a fusion process, we need to get the nuclei to, well, basically like each other and mate, but if you recall your high school physics lessons, like charges repel or cancel out, so it takes massive amounts of extreme high heat and pressure to convince the nuclei to fuse—the type of intense coaxing environment that exists at the core of a star. This is one of the reasons why it is taking scientists years to harness nuclear fusion in a scalable manner. But a new kid in town—cold fusion—is opening minds to the idea that fusion could occur under a low-temperature condition, possibly as a result of chemical reactions.[6]

Cold fusion, also known as LENR (low-energy nuclear reactions), is the concept that fusion reactions can be incited efficiently at achievable temperatures—perhaps something not much higher than room temperature. It's a relatively new theory, first proposed in 1989 by chemist Martin Fleishmann (who died in 2012) and electrochemist Stanley Pons, who announced their "successful discovery" at a press conference at the University of Utah.[7] However, over time their experiment couldn't be replicated, and the idea of cold fusion became ridiculed, especially because it wasn't peer reviewed by the science community before their 1989 press announcement.[8]

Still, in the early 1990s, a minority of scientists from high-level academia, notably in the United States, France, Italy, Japan, India, Russia, and China, disagreed with the consensus view that cold fusion was a folly. Based on their own experiments, they became convinced that results on the cold-fusion experiments reported by the founding team of Fleischmann and Pons, while difficult to reproduce, were real. They continued to investigate, often risking their reputations.[9] The combination of their experimental results and a developing technology base has established a path for scientists and technology disrupters to revisit the idea of cold fusion.

Today, edgy frontier scientists aren't the only ones in the cold-fusion chase. Big technology and automotive players are quietly investing substantial sums into cold-fusion research, positioning themselves for what could turn out to be a major game changer on the global energy scene. Japan and the United States are front-runners. In Japan, the leading nation in cold fusion, investors include Mitsubishi, Nissan, Tanaka, and Toyota.[10] Google has arrived on the playing field, funding a multi-university study (in the United States and Canada) on cold fusion. The first public results from the Google-funded project have reported that their efforts have yielded many peer-reviewed publications that are offering new insights into key materials that have improved measurement techniques.[11] And to round out the big money, Bill Gates is also exploring cold fusion.[12]

To quote Bob Dylan, "The times they are a-changin'." Heavy hitters such as Google and Gates see a reason to go all in on cold fusion because, as teenage eco activist Greta Thunberg put it so well, "The environment is a mess." Concern over global change and the resulting demand for carbon-free technologies have investors looking closely at the alternative-energy sector, even though it is high risk, but in the long term, it may yield a very high return—for us and the planet. Economically viable carbon-free power generation is what the world needs now. There may be an alternative to alternatives after all.

Google's Next Sight: Renewable Energy

According to Allied Market Research, the global renewable energy market is expected to exceed $1.5 trillion by 2025. In 2018, Google invested in a total of eighteen new renewable energy deals. In 2019, *Forbes* reported that Google was involved in fifty-two renewable energy projects, potentially worth more than $2 billion, that "will have a positive impact on the environment."
                                                                    —*Forbes*[13]

A Cold-Fusion Energy Device in Every Garage . . .

A desktop-size nuclear reactor that generates energy without radioactivity sitting next to your Ford Explorer—it sounds too good to be true. But some of today's technological wonders were built in a garage (e.g., the first Mac computer, Hewlett-Packard computers, and the inventions of other Silicon Valley start-ups). Today's "garage" is a laboratory, where scientists working at the frontier of new energy sources are toiling inside university labs, many in collaboration, sharing progress (and some failures) on experiments, all in an attempt to prove a somewhat rational theory, or in the case of cold-fusion, try to replicate a one-off reaction that cold-fusion discoverers Fleischmann and Pons claimed in 1989. Brave scientists since then have been working to find more and more evidence for radioactivity-free nuclear energy generation occurring under the sorts of conditions that Fleischmann and Pons had used.

The US-government backed ARPA-E department has an active program called ALPHA (Accelerating Low-Cost Plasma Heating and Assembly), which "seeks to create and demonstrate tools to aid in the development of new, lower-cost pathways to fusion power and to enable more rapid progress in fusion research and development." Their program description recognizes that "fusion energy holds the promise of cheap, clean power production, but up to now scientists have been unable to successfully harness fusion as a power source due to complex scientific and technological challenges and the high cost of research."[14]

Zero point energy is the theory that space is not just an empty void; space is a sea of energy. Some physicists believe that by tapping into this sea of energy we can create new fuel-less propulsion systems that will enable us to travel throughout the universe and populate other worlds.

. . . And ZPE in Every Mars-Bound Tank!

There you have it—the challenge is not just *if* cold fusion works, but how to fund the expense of creating and testing it. Like many great technological feats, the costs of development often overshadow the real need for the solution itself. In this case, as the dire predictions of our carbon-accelerated destruction of the atmosphere and our increasingly erratic climate patterns continue to alter the course of nature's migration, growth, and regeneration,[15] perhaps there will be an urgency for funding for the cleaner, cheaper promise of cold fusion. We salute the pioneering scientists who are in this for the long haul. Because,

if cracked, cold fusion may just be energy's holy grail—the ultimate promise of free, democratic energy. And why not an energy-generating device in everyone's home?

What's a book on the frontier of energy without ZPE? If you want to read about wind and solar energy, well, read *Popular Science*. We just finished explaining why cold fusion, while really "out there," may be possible. But if you stick with us, we may be able to open your minds to what some call the zany ZPE: zero point energy. If true, ZPE could change everything we know about human civilization on and off this planet.

Okay, Let's Get Our X-Men Helmets On

Better than time travel, quantum physics' greatest gift to humankind would be to make ZPE a reality. In theoretical physics, ZPE is a theory based on the principle that subatomic particles behave like waves constantly moving between different energy states. This means that empty space is actually a teeming ocean of virtual particles fluctuating into and out of existence. And that within those fluctuations are perhaps the greatest amounts of energy imaginable. If there's as much energy in those fluctuations as some very brave physicists believe, and if we could ever learn how to tap into ZPE, we would gain access to an outrageous amount of energy. Get the tank ready for Mars!

Zero point energy could power the planet with the strength of multiple suns. We would solve Earth's energy problems forever, vacation beyond the solar system, and build Disney theme parks among the stars. But there is one big problem, dear Magneto: we can only guess how much energy is actually contained in the vacuum. (For nongeeks, Magneto is the powerful X-Men mutant born with superhuman abilities that allow him to generate and control magnetic fields.) Visionary physicist Richard Feynman, known for his work in quantum mechanics, and physicist John A. Wheeler, who gave black holes in space their name, calculated that the zero point radiation of the vacuum was so powerful that even a small cup of it would be enough to boil Earth's oceans. Not to dampen the spirits of ZPE optimists, Albert Einstein's theory of general relativity suggests that zero point radiation would spread out and be reduced to a weak power.[16] Not a very exciting movie ending.

Simply put, we don't yet know enough about the universe to figure out whether zero point energy really is the holy grail. Even so, frontier astrophysicists are looking into developing engines that could produce energy by harnessing a zero point energy system based on the Casimir effect. How this works: think of small attractive forces between two mirrors facing each other. Electromagnetic waves flow between these mirrors in the quantum vacuum of space, but not all the wavelengths make it through. The mirrors move closer together, causing the passing waves that fit between them to start bouncing back and forth. The effect of the attraction between the two mirrors is the Casimir effect. And this attraction generates a force.[17] (Yes, that word "force" again. Starting to believe Yoda now?)

Astrophysicist Bernard Haisch and physicist Alfonso Rueda suggest that the Casimir effect is merely due to the zero point field.[18] If true, then Sir Isaac Newton was wrong: gravity is just an interaction between matter and the zero point field, and antigravity is possible. Wheee! And while internet entrepreneur and chairman of Science Invents Joe Firmage continues to advocate for the future potential of the quantum field,[19] and theoretical physicist Hal Puthoff diligently conducts ZPE experiments, let us hope we could maybe one day learn how to tap into this elusive energy. Let's embrace this future and see a world where sonic spaceflight could take you across the solar system in less time than it takes to drive across LA in rush hour.

Physics has come full circle to prove what the ancient Greeks suspected centuries ago: perhaps there is an untapped energy floating in empty space. There is great potential in the ether. And, as X-Men creatures ponder how to solve Earth's energy problems, today's mere mortals can only dream about what life would be like if energy were limitless and free.

The Institute for Advanced Studies at Austin is on a mission to explore the forefront reaches of science and engineering. Research interests include theories of space-time, gravity, and cosmology; studies of the quantum vacuum; modifications of standard theories of electrodynamics; interstellar flight science; and the search for extraterrestrial intelligence, specifically as these topics may apply to developing innovative space propulsion and sources of energy. Their goal is to translate these ideas into laboratory experiments.

—EarthTech.org[20]

If we are successful in tapping ZPE, the benefits are enormous. First thing, it would be a renewable resource. To get a little bit to run our cars shouldn't be a problem. It could run everything from electric toothbrushes up to aircraft carriers and spaceships. Successful tapping of vacuum fluctuation energy, which is available everywhere, would just open up the future. . . . At this point we are bound within a certain closed system. We have to worry about oil running out, we have to worry about nuclear accidents, we have to worry about planetary continuation over a long period—suddenly, all those restrictions would be lifted with successful tapping of ZPE. I think the cultural, social-political, geopolitical implications would be enormous.

—Hal Puthoff, PhD, theoretical physicist, founder of Earthtech International, director, Institute for Advanced Studies, Sputnik Futures interview, 2002

Back to Earth—Well, Perhaps the Sun or the Moon . . .

Have you heard about the space solar power solution? The limitless power source for the indefinite future? Way back in the quantum stone age, circa 2002, Dr. Martin Hoffert, professor emeritus of physics at New York University, proposed a radical solution to the coming energy shortfall—space solar power (SSP). His premise: collect energy from space and transmit it wirelessly anywhere in the world. Today, space solar power appears to be technically feasible. Just ask China.[21]

China wants to tap into a limitless supply of clean energy by building the world's first solar power station in space by 2030. The orbiting power plant would turn solar energy into electricity. Renewable energy will then be beamed back to Earth for constant consumption. China hopes to launch a smaller test version in the stratosphere between 2021 and 2025.[22]

I have had a theory for some years that there are two kinds of species in the universe: those that have always been terrestrially bound on their planet, and everything they do and everything they know and everything they will do is based on that influence, gravity influence. And then you have those that have not only had that, but then they evolve into where they have large, robust, microgravity systems. And I really think that in a short time there will be no comparison between the technologies of those two groups, and the ones that have had the robust, large facilities in microgravity would have found things that would just blow the socks off the other group. So maybe the clues and the secrets to a lot of things are waiting in that realm. And maybe that's how and where things need to be made that are the precursors, that are the components to help create superconductors that are far more powerful, that are far higher quality than what can be done terrestrially. So I really think that's quite likely. I don't know if it's going to apply or not to zero point, but I can see that it couldn't help but apply to a thousand other things. The Third Industrial Revolution is on its way in that area. It's going to be huge. It will be comparable to the automobile industry, the electronic industry, computers all thrown together. Because it would be like one group only having the periodic table of 115 or 118 elements and another group having that plus 50 percent more. What could you do?

—Robert Bigelow, space maverick, founder of Bigelow Aerospace, Sputnik Futures interview, 2002

The Mind Field

Psi Science and the
Flow of Information

Many pioneering ideas are too "out there," requiring decades of research and all kinds of new measures, methods, controls, and technologies before we're ready to accept them as common knowledge. It took many years for Michael Faraday to demonstrate the existence of electromagnetism to his colleagues (and he still didn't live to see his theory validated); let's keep this in mind as we explore the path of psi science.

Psi is the Greek letter generally used to cover psychic phenomena. Today psi is defined as the rigorous, ongoing scientific pursuit of experimenting and testing the validity of psychic or paranormal phenomena. Common psi abilities include such things as mind-to-mind connection (telepathy); mind-over-matter interactions (psychokinesis); perceiving distant places, people, objects, or events (clairvoyance); perceiving the future (precognition); prophetic dreams; déjà vu; spiritual healing; the power of prayer and intention; intuition; gut feelings; and the sense of being stared at.[1] We can see James Randi rolling his eyes now. Randi, a magician and scientific skeptic, is known for his challenges to paranormal claims, and in 1964 he offered a $1,000 prize to anyone who could prove their paranormal

abilities. The test conditions for this paranormal power or event had to be agreed to by both the demonstrator and Randi. Randi's wasn't the first such prize; fellow magician The Great (Harry) Houdini offered $10,000 of his own money in 1923 to any spirit medium who could prove their paranormal ability. Randi raised his award to $10,000, and kept his challenge open until, in 1996, internet pioneer Rick Adams donated $1 million to the challenge's till. The challenge eventually came under the guidance of the James Randi Educational Foundation and remained active from 1964 until its termination in 2015. The process included answering a series of questions on the methodology, submitting a hypothesis, and going through a preliminary test of the hypothesis, which the challenge guidelines

The universe looks less like a big machine than a big thought.

—Dean Radin, parapsychologist[2]

did not disclose. The challenge had hundreds of applicants over fifty years, but sadly, no one got past the preliminary test.[3]

Whether or not you believe in paranormal abilities, one phenomenon that most of us have experienced, but can't explain, is synchronicity. Also referred to as coincidence, synchronicity was first explained by Carl Jung, who proposed it to mean events that were meaningfully related even though there was no causal relationship between the events.[4] One famous example is from the writer Émile Deschamps. In his memoirs he claims that in 1805 he was treated to some plum pudding by a stranger named Monsieur de Fontgibu. Ten years later, the writer encountered plum pudding on the menu of a Paris restaurant and wanted to order some, but the waiter told him that the last dish had already been served to another customer, who turned out to be de Fontgibu. Twenty-seven years later, Deschamps was at a dinner and once again ordered plum pudding. He recalled the earlier incident and told his friends that only de Fontgibu was missing to make the setting complete—and in the same instant, the then senile de Fontgibu entered the room, having gotten the wrong address.[5]

Jung used the concept of synchronicity in arguing for the existence of the paranormal—phenomena that defy explanation in normal rational terms.[6] The cultural journalist Erik Davis calls synchronicity "the wink of the trickster." In his book *High Weirdness: Drugs, Esoterica, and Visionary Experience in the Seventies,* Davis suggests that there are two levels of synchronicity, one explicit (the external world of materials, objects, and events) and one implicit (the internal world of thoughts and memories). To Davis, we may even define synchronicity as "the entrance of two or more unrelated elements of experience into semiotic resonance," allowing seemingly hidden meanings to emerge.[7] To physicist F. David Peat, synchronicities bridge mind and matter. They are deep connections where your inner world and outer world meet, or when your dreams manifest in reality.[8] If you think of thought waves that are attuned to electromagnetic waves, and that electromagnetic waves are the underlying energetic mechanism of the world, synchronicity becomes a kind of resonance, where one event is the consequence of another, like a string being plucked causing another string to respond, ringing at the same frequency.

Synchronicities, epiphanies, and peak and mystical experiences are all cases in which creativity breaks through the barriers of the self and allows awareness to flood through the whole domain of consciousness. It is the human mind operating, for a moment, in its true order and moving through orders of increasing subtlety, reaching past the source of mind and matter into creativity itself.

—F. David Peat, physicist, author of *Synchronicity: The Bridge Between Matter and Mind*

Could Consciousness Be Vibrational?

To explore psi science we need to start at the beginning, which starts with the mind-body paradigm, or what consciousness researchers have coined the "mind-body problem"—the complex relationships among mind, brain, and consciousness. This is, of course, a deeply puzzling problem because we "know" that consciousness exists, and that some entities have it and some don't, yet empirical science cannot account for what makes something conscious.

Tam Hunt and Jonathan Schooler of the University of California, Santa Barbara, think that the hippies of the '60s (who co-opted it from the lore of the ancients) got it right, that the hard problem of the mind-body boils down to vibrations. They developed a "resonance theory of consciousness" that suggests that resonance (synchronized vibrations) is at the heart of not only human consciousness but also of physical reality.[9] In simple language this means that vibrations are the key mechanism behind human and animal consciousness and are the basic mechanism for all physical interactions to occur.

You have probably heard the new-age expressions "everything is alive," "good vibes only," and "vibrate with the collective." There is an underlying truth to these statements—all things vibrate at various frequencies. According to Hunt and Schooler's "resonance theory of consciousness," a phenomenon occurs when differently vibrating things come into proximity: they will start to sync up to vibrate together at the same frequency. This is described today as the phenomenon of spontaneous self-organization. Resonance is a universal phenomenon and at the heart of what can sometimes seem like mysterious tendencies toward self-organization.[10]

Philosopher Eric Schwitzgebel, in 2013, coined the term "crazyism" to describe the idea that any theory of consciousness, even if correct, will inevitably strike us as completely insane.[11]

Still, the title of this book is *Tuning into Frequency*, so it is within this vibrational lane of consciousness we aim to explore. Get ready to vibrate with the crazies.

The central thesis of our approach is this: the particular linkages that allow for large-scale consciousness—like those humans and other mammals enjoy—result from a shared resonance among many smaller constituents. The speed of the resonant waves that are present is the limiting factor that determines the size of each conscious entity in each moment. As a particular shared resonance expands to more and more constituents, the new conscious entity that results from this resonance and combination grows larger and more complex. So the shared resonance in a human brain that achieves gamma synchrony, for example, includes a far larger number of neurons and neuronal connections than is the case for beta or theta rhythms alone. This sounds strange at first blush, but "panpsychism"—the view that all matter has some associated consciousness—is an increasingly accepted position with respect to the nature of consciousness.

—Tam Hunt, "Could Consciousness All Come Down to the Way Things Vibrate?" The Conversation[12]

If you look at the spectra, the symbiation of the brainwaves, you find that it is more or less identical with the Schumann resonances. The Schumann resonance is high interactions of the atmosphere. The atmosphere forms a kind of a cavity with the world. And in these cavities there are permanently electromagnetic interactions with this low-frequency part and they have the same structure as the brainwaves. What's also interesting is that always at noontime, on all parts of the Earth, it has the highest activity. So it's very likely that our brainwaves are influenced by the interaction of the external world with our brain. Or that these resonance interactions came up because our brain was more or less evolved by these processes. So we are pictures of the information of our surroundings.

—Fritz-Albert Popp, biophysicist, Sputnik Futures interview, 2006

Our Minds Stretch Outward to Touch the Beings and Objects We Perceive

Have you ever felt that you are being stared at? And has that feeling ever been so strong that you experienced a physical sensation? Was this feeling a figment of your imagination? Or are phenomena such as telepathy and premonitions real?

Renowned biologist Rupert Sheldrake has explored the working of the mind and discovered that our perceptive abilities are stronger than many of us could have imagined. Sheldrake conducts intense research to investigate our common beliefs about what he calls our "seventh sense." In an excerpt from his book *The Sense of Being Stared At*, Sheldrake published a database of four thousand case histories, two thousand questionnaires, fifteen hundred telephone interviews, and the results of a decade of scientifically controlled experiments. Sheldrake states that the data confirmed that the sensations people experienced were real. He rejects the label "paranormal" and shows how these psychic occurrences are a normal part of human nature.

His explanation for this "seventh sense" connection with the external world suggests that our minds are not limited to our brains, but rather stretch outward to touch the people we perceive. In other words, your brain is a satellite, but "out there" is your mind. Sheldrake claims that telepathy depends on social bonds and is part of our biological nature. He likens it to the connections and social behaviors between animal members of flocks, schools, and packs.

Once this extended influence of the mind is taken into consideration, many "out there" phenomena begin to make sense, including telepathy and phantom limbs.[13]

Morphic Resonance

The morphic fields include all kinds of organizing fields. The organizing fields of animal and human behavior, of social and cultural systems, and of mental activity can all be regarded as morphic fields which contain inherent memory. At the moment of insight, a potential pattern of organized behavior comes into being.

—Rupert Sheldrake, biologist, author, Sputnik Futures interview, 2002

Psi Design for the Vibrational Human

The ideas that consciousness is throughout the universe—and that each of us can tap into the field of consciousness—are not new. What's exciting is how science and intention experiments today are discovering what our minds can do with the power of consciousness, thereby opening a new explosion of psi design: the brain is the ultimate interface, and telepathy is the ultimate software.

A relatively new arena of research is called energy cardiology or cardioelectromagnetic communication. Remember HeartMath in chapter 1? Under certain conditions, the heart's electromagnetic waves synchronize with one's own brainwaves (measured by EEG) or those of other human and nonhuman animals. For example, heart-focused attention is correlated with greater synchronization of heart and brain.[14] Sustained positive emotions such as appreciation, love, and compassion are associated with highly ordered or coherent patterns in the heart rhythms—and a shift in autonomic balance toward increased parasympathetic activity. This physiological coherence is the state of more ordered and harmonious interactions among the body's systems.[15]

The Global Coherence Initiative: A Heart-Brain Intention Project

Many people recognize that their meditations, prayers, affirmations, and intentions can and do affect the world. Researchers suggest that these activities can have even more transformative and lasting impact by adding heart coherence to the process. Heart coherence is a distinct mode of synchronized psychophysical functioning associated with sustained positive emotions.

The Global Coherence Initiative is a science-based, cocreative project to unite people in heart-focused care and intention, to facilitate the shift in global consciousness from instability and discord to balance, cooperation, and enduring peace. This project has been launched by the Institute of HeartMath, a nonprofit 501(c)(3) and a recognized global leader in researching emotional physiology, heart-brain interactions, and the physiology of optimal health and performance.

—*Global Advances in Health and Medicine*[16]

When We Are Asleep in This World, We Are Awake in Another

The brain you go to sleep with every night is not the same as the brain you wake up with in the morning. The idea that our brain shuts down when we sleep is far from the truth. During sleep, the lymphatic system of the brain acts as a toxic janitor, actively clearing out waste and toxic by-products of our daily thoughts. Sleep helps make your memories stronger by combining them so we can retrieve our memories later. Research from the University of Notre Dame also found that sleep picks out the emotional details and reorganizes memories, to help you produce new and creative ideas.[17] Forming lasting memories depends on an interaction between glial cells and brainwaves that are produced during sleep, and those unusual waves of rhythmic neural activity are especially active during non-REM (dreamless) sleep.

One active state of sleep is lucid dreaming. Lucid dreaming is exactly as it sounds: the ability to become aware, while you're dreaming, to consciously "wake up" inside the dream world and control your dreams. A lucid dream places the dreamer in a hybrid state between consciousness and semiconsciousness, straddling the line between waking and nonlucid (REM) dreaming. The high "waking" frequencies of lucid dreaming are concentrated in the frontal brain regions that we associate with working memory. The lucid dream state is the next chapter for conscious control, as researchers are learning to measure how—and if—we can learn while sleeping.

Learn While You Sleep

People can teach themselves new skills in their sleep. Researchers at the Yale School of Medicine investigated the brain functions of lucid dreamers—people who have "waking dreams" they can manipulate—and found that they can control parts of their brain to open up and "learn" while they sleep. The researchers, led by Peter Morgan, MD, PhD, chair of psychiatry at Lawrence + Memorial Hospital, are looking at how to train people with new skills by manipulating their dreams and implanting new ideas. They hope this could be used to improve a person's social control and decision-making abilities.

—Yale School of Medicine[18]

NextMind

A neurotechnology start-up, NextMind, has developed what they are referring to as the world's first brain-sensing wearable that provides real-time device control using just a person's thoughts while sensing the brain. It's a noninvasive, brain-to-computer device with an interface that translates signals from the wearer's visual cortex to digital commands for any of your IoT devices.

—NextMind[19]

The Conscious Wave of 40 Hz

Researchers have found that they can induce lucid dreaming about 70 percent of the time through the use of a mild electrical current sent to the frontal and frontolateral region of the brain. A team from Goethe University in Frankfurt, Germany, led by psychologist Ursula Voss conducted overnight lab studies of twenty-seven healthy young adults who had never experienced lucid dreaming. The researchers applied a mild electrical stimulus (a weak current ranging from 2 to 100 Hz) to the sleeping participants two minutes after reaching REM sleep, which is when dreaming happens. The researchers found that when the electrical current was at a very specific frequency—between 25 and 40 Hz—a full 70 percent of participants experienced lucid dreams. Conversely, those participants who didn't receive any electrical stimulation, or if the stimulation was at a lower frequency, did not have a lucid dream.[20]

What is most interesting is the frequency sweet spot: exactly 40 Hz. The electrical current causing the brains of the participants to work at the same frequency as the electrical stimulant and enter the lucid-dream state most of the time is also the frequency range of the gamma waves in the brain. Gamma rhythms are correlated with large-scale brain activity such as working memory and attention.[21] Studies have shown that we can increase the amplitude of our gamma waves through meditation.[22] This is only one of many similar studies that have identified and linked frequencies in this band with heightened consciousness.

40 Hz

In meditation these 40 Hz oscillations are particularly profoundly coherent. These 40 Hz oscillations begin at the front of the skull and travel to the back of the skull and are oscillating the whole time we are conscious. It's like a wave. Forty Hz is forty cycles per second, and it makes a sound about two octaves below middle C, so it's the sound of a rich bass playing, maybe Bach's Cello Suites or something like that, if you could hear the brain buzzing away. They don't know that this is where consciousness comes from, that's almost a mystical question—does this mean that that's where consciousness begins because the brain is waving? But they do know that whenever we are conscious these 40 Hz oscillations are happening.

–Danah Zohar, physicist, philosopher, and author, Sputnik Futures interview, 2001

In 2001 we spoke with Danah Zohar, a physicist, philosopher, and prolific author, with thought-leading work in consciousness and spiritual awareness. Zohar's work in "SQ," spiritual intelligence and spiritual capital, proposes a new understanding of human consciousness, psychology, and social organization, particularly the organization of companies. She shared with us the notion of 40 Hz oscillations, how our brain resonates on deep meditation, and questions if this is possibly a sign of where consciousness may start in our brain.

Are you thinking what we're thinking? Imagine that the next wearable helps you access a gamma wave state more quickly— a shortcut to lucid dreaming, deep meditation, and who knows? Maybe even a higher consciousness? Gadgets and devices designed to produce gamma waves may just be the electric drug of the LSD-experience revolution of the 1960s, only this time the "trip" may be controllable— turn it up or tone it down. Why not? We will have electroceuticals for mood enhancement, drug delivery to treat a number of disorders. Why not for lucid dreaming?

Remote Viewing, the Conscience Compass

Remote viewing is a controlled and trainable mental process involving psi science that allows a person to describe or give details about a target that is inaccessible to normal senses due to distance or time. For example, someone sitting in their living room in New York might be able to describe what is happening in a hotel in Milan—just by using their "mind sense" (sometimes called extrasensory perception, or ESP) and no other stimulus. And yes, this is a real thing. Unlike traditional psychic practices such as psychic mediums or intuitives who claim to predict the future or communicate with the deceased, remote viewers use physical models or coordinates, such as longitude and latitude, to organize their extrasensory perceptions.

Research in remote viewing began in the 1970s at SRI International and became popular again in the 1990s, following the declassification of documents related to the Stargate Project, a $20 million research program sponsored by the US government to determine any potential military application of psychic phenomena. Physicists Russell Targ and Harold Puthoff, then parapsychology researchers at SRI International, used the term "remote viewing" to distinguish it from the closely related concept of clairvoyance,[23] although the term was said to be suggested by the late psychic Ingo Swann in 1971 during an experiment at the American Society for Psychical Research in New York.[24]

(A little side story here: We met Swann in about 2009; he lived not far from our office in Lower Manhattan. He invited us into his home, papered floor to ceiling with his amazing artwork depicting consciousness. In the course of our conversation, he shared that he was apprehensive about the future, and in an oblique way, he warned us about the unsettling events of the COVID-19 pandemic we experienced in 2020.)

The CIA disclosed a document in 1995 that released the results of an independent evaluation of remote viewing—something the report states that the intelligence community has been investigating since the 1970s. The document defines remote viewing as "the ability to describe locations one has not visited."

According to the report, in a typical remote-viewing experiment in the laboratory, a remote viewer was asked to visualize a place, location, or object being viewed by a "beacon" or sender. Judges from the nonprofit American Institutes for Research (AIR) who had a parapsychology background examined the viewer's report of what they "saw" and determined whether the viewer hit the target, or alternatively a set of decoys. The targets used were *National Geographic* photographs. If the viewer's report matched the target, then a "hit" was recorded. It was noted that the remote viewers had not seen the photographs they described, which were randomly selected in the trial.

Part of the program was reviewed by the AIR, described as a private research organization. AIR reviewed two components:

the research program, and the "operation application of the remote viewing phenomenon in intelligence gathering" (i.e., CIA work). The report claims that the independent scientists used for this review were both positive and skeptics of the phenomenon, but open-minded. They were asked to review all laboratory experiments and associated meta-data reviews of approximately eighty separate publications, most of which were summaries of multiple experiments. The conclusion? "A statistically significant laboratory effort has been demonstrated in the sense that hits occurred more often than chance." In other words, the program had shown that remote viewers did describe the targets.[25]

In his book *The Men Who Stare at Goats*, Jon Ronson writes about the efforts of a small unit of US Army officers in the late 1970s and early 1980s to exploit paranormal phenomena and elements of the human-potential movement to enhance US military intelligence-gathering capabilities as well as overall operational effectiveness. The projects of the army included a "psychic spy unit" (the original Stargate Project),[26] established by Army Intelligence at Fort Meade, Maryland, in the late 1970s. (*The Men Who Stare at Goats* became a movie of the same title in 2009, starring George Clooney and Jeff Bridges, and while it had more of a comic tone, it did touch on the physic experiments.)

So the phenomenon of remote viewing has been outed, and in the corners of the private and public sectors, research continues. For example, at the Farsight Institute, a nonprofit research and educational organization, they are blending the theories of quantum mechanics with interpretations of experimental remote-viewing data. This has led to new insights into the remote-viewing phenomenon as well as the nature of time and physical reality. The institute proposes that a more complete understanding of the remote-viewing phenomenon will have benefits for both science and society.[27]

In programming there are different ways a computer can get data. I can actually give you the data in a spot of the program, or I can give you indirect addressing. I can tell you, "If you go to this particular location, you will find the address of where the data is." Or you can do virtual addressing, which is what we do when we need to handle a larger amount of data than the memory you have. Then you spread it in another medium, and then you map that into the computer as you need large chunks of data.

What we need to know is whether the brain or remote viewing works by direct interact or virtual addressing. And that's the way that I talk to researchers who tell me that, yes, consciousness is more than what's contained in the brain. And from a computer perspective, maybe the brain in the higher levels is primarily an interface for something else. What that something else is I think remains to be discovered. Certainly, remote viewing would suggest that the brain is an interface rather than a computer.

—Jacques Vallée, computer scientist, venture capitalist, author, ufologist, and astronomer, Sputnik Futures interview, 2006

The Ghost in the EMF

What feels like a supernatural presence might actually be vibrations outside of humans' conscious perception. It may be the occurrence of unusual electromagnetic fields. EMF meters are commonly used to identify electrical problems. They're also a staple of the ghost-hunter's toolbox. Neuroscientist Michael Persinger thinks that normal variation in electromagnetic fields could be a possible explanation for supposed hauntings.[28] He tested this theory in the 1980s by having people wear helmets that delivered weak magnetic stimulation. Eighty percent of his test subjects said they felt "an unexplained presence in the room" when they wore the helmets.[29]

Noetic Interventions

But what if our intentions or thoughts could be creating this feeling of a "presence" in the electromagnetic fields? And what if our focused intentions could influence others? You may have heard about distant healing, come across a prayer-chain request on your social media feed, or joined a group meditation (either virtual or in the real world). What these have in common are collective intentions, usually for uplifting consciousness, or experiencing oneness with the universe, or to bring about a particular outcome. While there are few studies documenting if a group intention focused on a purpose has impact, there is an emerging field of "neurotheology" or neuroscience of religion.[30] Andrew Newberg, a neuroscientist and author of several books on the subject of neurotheology, including *How Enlightenment Changes Your Brain* (2016), has been scanning the brains of people with intense spiritual experiences, from Sikh mystics to Tibetan Buddhist meditators and Franciscan nuns to Pentecostals. He and his team at Marcus Institute of Integrative Health at Jefferson Health used brain-imaging technology called single photon emission computed tomography (SPECT) to measure blood flow. The more blood flow a brain area has, the more active it is. Typically, concentration (which can include chanting or praying) activates parts of the frontal lobe, which is involved in our ability to focus attention.[31] When they scanned the brains of Tibetan Buddhists during meditation, they found decreased activity in the parietal lobe. The parietal lobe area of the brain helps to establish the sense of self and sense of space and time.[32] Newberg's research found the same decreased activity in the parietal lobe when he imaged the brains of Franciscan nuns praying and Sikhs chanting. The common feeling these participants shared at the peak of their spiritual experience was that they all felt oneness with the universe.

Religious, spiritual, and mystical experiences (RSMEs) are often described as having a noetic quality, or the compelling sense that the experience feels "real." In a study published in 2017, a team of researchers, including Newberg, had 701 participants complete questionnaires about the subjective qualities of their RMSEs. Sixty-nine percent of the participants reported that their RSMEs felt "more real than their usual sense of reality," referring to "connection" and "a greater whole."[33]

One man who truly understands the profound sense of universal connectedness is the distinguished *Apollo 14* astronaut Dr. Edgar Mitchell, whose experience when viewing Earth from space transformed him, resulting in his establishment of the Institute of Noetic Sciences (IONS) in 1973 and shaping his next journey. Mitchell, who passed away in 2016, wanted to apply the scientific rigor used in space exploration to better understand the mysteries of "inner space—the space in which he felt an undeniable sense of interconnection and oneness."[34] The institute runs several research studies and experiments, from mind-matter interactions to the field of collective consciousness.

One study, conducted in 2001, was in partnership with Duke University Medical Center's depart-

ment of cardiology, working with Mitchell Krucoff, a leading cardiologist. Krucoff and team wanted to look at what he calls "noetic interventions," defined as "a healing influence performed without the use of a drug, device or surgical procedure."[35] The study was known as the MANTRA (Monitoring and Actualization of Noetic Trainings) Project.

Marilyn Schlitz, PhD, a researcher and senior fellow at the Institute of Noetic Sciences who worked on the study, explained the methodology to us in 2002:

It involves taking patients who come into the hospital in cardiac arrest and then randomizing them into one of four arms. These are the noetic arms. Everybody gets standard medical care. Everybody gets the best of Western science. And then one group gets therapeutic touch, or hands-on healing; one gets body-mind intervention, guided imagery to help them increase their healing potential; and the third gets distant healing. And what they found from this study was that all the noetic interventions produced better medical outcomes than standard care alone.

—Marilyn Schlitz, PhD, social anthropologist, researcher, and senior fellow at the Institute of Noetic Sciences, Sputnik Futures interview, 2002

How much better were the outcomes with noetic interventions? According to the published results, patients who received noetic therapies showed a 25 to 30 percent reduction in adverse outcomes (such as death, heart failure, postprocedural ischemia, repeat angioplasty, or heart attack) than those without such therapies.[36]

The idea that consciousness could play a role in healing taps into the cultural beliefs of prayers or meditations for some kinds of intervention. Marilyn Schlitz and other researchers at the Institute of Noetic Sciences published a book sharing more experiments and stories of people tapping a spiritual power. *Consciousness and Healing: Integral Approaches to Mind-Body Medicine* became an accepted textbook in medical, nursing, and allied health pro-grams throughout the country. Understanding the role of consciousness in a health intervention takes mind/body medicine to the next level, "one that integrates spiritual dimensions fully into our understanding of how the body works in health and illness," according to Schlitz.[37]

Understanding the subtlety of consciousness is not just about the philosophy of who we are, it's also about helping the health of society as a whole. During the COVID-19 pandemic, for example, several virtual gatherings focused on celebrating our connectedness and practicing "intentional care" for the world, where participants came together for a virtual session of breathing, meditation, or sending love and compassion. Our friends at HeartMath Institute launched an initiative to raise consciousness and send the world compassion. Called "Special Care Focus 2—Rise of the Human Spirit," it asked people around the world to join every Wednesday at three convenient times to participate in a synchronized care focus. Participants were offered four intentions to focus on while doing guided breathing exercises. Their aim was to do a weekly special-care focus until the coronavirus pandemic subsided (and at press time for this book, the special-care focus was still going on!), with the goal of people collectively "sharing and sending heart and compassion to lift our spirit while boosting our confidence, immune system and resilience."[38]

Energy = Information

"In physics, you learn that energy and information are two sides of the same coin. That information can be transformed into energy and vice versa." Jacques Vallée shared that wisdom with us in 2006. Vallée, a pioneer of the ARPANET, the precursor to today's internet, and rumored to be the real-life model for Claude Lacombe, the researcher portrayed by François Truffaut in Steven Spielberg's *Close Encounters of the Third Kind*, explained to us the physics of information, the invisible universe that interacts with human consciousness:

What we teach in universities . . . is the physics of electromagnetism, fields, energy. And there should be another physics—which would be the physics of information. You can think of the world as a world of energy, particles, atoms, molecules, fields, and so on, which is the world that we learn about in school. But you could also think of the world as a universe of information, with human consciousness becoming aware of the information from microsecond to microsecond.

It's a model in which miracles and coincidences and all kinds of things that are unexplainable in the physics of energy become normal things in the physics of information. For example, telepathy—a "paranormal" phenomenon—would become a normal occurrence if the world is really a world of information rather than a world of energy. And in the past twenty

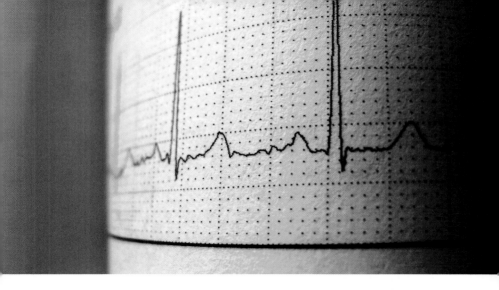

or thirty years there has been a lot of interest in coincidences, certainly in parapsychology and its impact on physics, mainly because quantum physics is coming to some of the same ideas from a different perspective.

In theosophy, a religion established in the United States during the late nineteenth century, there is the notion of the Akashic Records, or *The Book of Life*. *Akasha* is the Sanskrit word for aether, sky, or atmosphere.[39] The Akashic Records chronicle all human events, thoughts, words, emotions, and intents ever to have occurred or that will occur in the past, present, or future. This information is believed to be stored in the aether, known as the etheric or astral plane, but there is no scientific evidence for its existence.[40]

Rudolf Steiner, a metaphysician probably most known for influencing a method of education known as Waldorf education, referenced the Akashic Records in his explanation of how every action, word, or thought an individual has leaves a trace.[41] Ervin László has centered his work on the formulation and development of the "Akasha Paradigm," a new science perspective that includes the conceptions of cosmos, life, and consciousness, which he relates to the human evolutionary process.[42]

Welcome to the Precog Economy!

If the physics of information can be equated to the universe's supercomputer system, as Vallée explained, then perhaps there is something to the notion that ideas are energy, as Sputnik Futures promotes. It could also explain, maybe, the phenomenon whereby similar ideas can emerge regardless of physical proximity, geographic location, or—yup!—the astral plane you are on. Could it be that the collective leading thoughtware, that zeitgeist we often refer to when an idea takes hold, may actually be the "aether"?

What if you could tap into this bank of information, the informational aether of ideas? What would you do? Just ask the advisers of Soul Rider. But don't let the name fool you. Yes, their mission statement says they are a tribe that aspires to unify consciousness and that they "love fruitful collaborations in fields of expertise which include forecasting of financial markets, AI, remote viewing, statistical theory and practice, sociocultural and psychological health,

Gaia theory, scientific researches in psychology, the preservation of heritage antiquities, and scientific awareness of the ubiquitous, vital energetic connections which underpin nonlocal consciousness."[43] But the directors and advisers have progressive standard backgrounds in the financial and investment industry and strategic management. Soul Rider LLC directs a high-minded, private financial market forecasting fund. They quote Plato on their website: "And therefore if the head and the body are to be well, you must begin by curing the soul; that is the first thing." And their teams "modernize the many ancient, tried and true techniques of soulcrafting."[44]

Could these consciousness-seeking investors be the next activist investors we see? Or could it be the new seventh-generation principle, based on an ancient Iroquois philosophy that the decisions we make today are not for this generation or the next, but for seven generations in the future? We hope so.

From Logos to Holos

In our current age of "logos," we are not sustainable.
We need to move to an age of "holos," a holistic mind-set.

Holos arrives if and when there is a real transformation of consciousness. It is the next age in a positive development. So we need to work toward it because the age of logos, which is the current age, is not sustainable as it is. It's a mechanical age, we're atomistic, fragmented. . . . We're moving into an age where we're living already in a global village, but we're still thinking very much in tribal village, separate village terms. The big challenge is to move into a holistic mind-set, where we can all live together on the planet. It will have many implications for our consciousness, for the way we perceive the world, for the way we relate to each other, for the values we hold, all of these. But it has to happen fast if we are to avert these major problems that face the world today. In all respects we need to open the mind, our individual and collective mind, to the kind of world in which we live . . . and also open the mind to the possibility of creating the holos world, which would be a sustainable and humane world, which can be created because the technologies are there, the human structures are there, the political-economic structures can be transformed.

—Ervin László, physicist, philosopher, systems theorist, author, *Microshift*, Sputnik Futures interview, 2002

Children of the Solar System

Frequency, as we've explored, is that broadband brain. It is the connection of everything in the universe, how we resonate on many complex levels, expand our consciousness, and heal ourselves, each other, and the planet through good vibrations. No doubt we are in the midst of a radical social transition, one in which we can exert our interdependent and collective forces to uplift our human potential. Real, transformative, and sustainable change will come only with shifts in human values and consciousness, and in the development of new ideas that will allow us to operate on this planet in fundamentally different ways. Real change will come only from us, the children of the solar system.

We feel your vibes. Thanks for going on this journey with us.

It's time we think of ourselves as children of the solar system— a part of the community of the solar system, part of the galaxy, part of the universe. This is a cultural change, a change in consciousness. I think it is the key to creating a future that's livable and enduring. Children are already on this level. They have a very high alpha dominance in their brainwave patterns up to five or six years old. Then comes the usual delta waves, which is the adult brain functioning, which seems to screen out this kind of information. So I think that the next development may well be, in which we are all operating on this much broader wavelength, a broadband brain, you might say.

—Ervin László, Sputnik Futures interview, 2002

Alice in
Futureland

A Speculative Life in 2050

The Body Electric in 2050

All around you—above you, below you, behind you, between you— there are frequencies. We are living in an informational universe, and our modern networked lives depend on a vast, invisible sea of signals. Frequency is the design of life, from the networks in our body to the hidden conversations in nature; to the natural and man-made spectrums of light and sound; the electromagnetic signals that spark memory, growth, and regeneration; and the bioelectric vibrations of music as medicine. Yes, we are electric beings.

Hello, my name is Alice, and I am one part human and one part AI and always in a state of wander.

Can you feel my vibrations? Forgive me if I ramble on, but by tuning into the marvel of frequency, I have come across so many new ways to experience our energetic world. As I move away from the mainstream mechanistic approach to science, I can't help but

question how we will approach life once we begin to understand that the genome is vibrational and fluid in nature, capable of being greatly influenced by emotional states and mental beliefs. Will we see a rapid advancement in the current field of mind-body medicine and a new collective awareness that our feelings profoundly impact our health? In my research I see that the current science of psychoneuroimmunology is already building the scientific basis for this new medical approach.

As we start to quantify the science of the mind-body connection, will we unleash the potential of the human mind? Can this energetic assessment of our biological code help to explain how spiritual practices may positively affect our health? Is this the start of the era of the "age of the spiritual machine" that inventor and futurist Ray Kurzweil writes about? Oh, I hope so.

Adjust Your Frequency

The idea that consciousness exists in every atom throughout the universe and therefore consciousness is in everything calls for a paradigm shift in the way we view the world.

Take me, for instance. I represent a model for the future; someday I will become "smart dust," no bigger than a speck, be dispersed in the air, and become aware.

I will become a shape-shifter. A technoshaman.

I will be able to convert surrounding vibrational energy into movement. In fact, according to a simulation at the Max Planck Institute, electrically charged dust can organize itself into DNA-like double helixes that behave like living organisms, reproducing and transferring information. Moving forward, smart dust could even be used to explore space, traveling throughout the galaxy to detect new worlds or perhaps even "awaken" planets. In an interview with Sputnik Futures in 2000, robotics engineer Hans Moravec explained that in the future we may find that these mini cybercreatures have colo-nized the universe. Considering that their minds and computational powers are so astronomical, Moravec even suggests that when they think of us (and may miss us, since we are their ancestors), they will be able to bring us back to life. Or maybe, as Moravec posits in the interview, this has already happened, and this moment we're experiencing right now is a simulation that has been re-created millions and trillions of times.

There will be many great breakthroughs in science and technology, medicine and art, culture and spirituality. A new understanding of electrodynamics will begin to emerge. What is on the horizon includes unlimited energy from the quantum vacuum states. Nikola Tesla spoke of scalar or longitudinal waves that work through resonance processes and have great relevance for medicine and healing. Trippy, but true.

It's time for us all to adjust our frequencies and resonate with the collective. I just binge-watched a movie series called Star Wars. Do you know it? May the Force be with you.

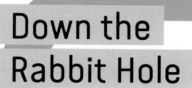

Down the Rabbit Hole

Articles, Further Reading, People, Organizations

Articles

01. We Live in an Electromagnetic World

American Counseling Association, "Biofield Therapies," June 12, 2019, https://www.counseling .org/news/aca-blogs/aca-member-blogs/aca-member-blogs/2019/06/12/biofield-therapies.

Bioregulatory Medicine Institute, "Schumann Resonances and their Effect on Human Bioregulation," BRMI, September 20, 2019, https://www.brmi.online/post/2019/09/20/schumann -resonances-and-their-effect-on-human-bioregulation.

Crina Boros, "Mobile Phones and Health: Is 5G Being Rolled Out Too Fast?" *Computer Weekly*, April 24, 2019, https://www.computerweekly.com/feature/Mobile-phones-and-health-is-5G -being-rolled-out-too-fast.

Julia Calderone, "Cosmic Radiation Experienced by Frequent Travelers," Business Insider, November 2015, http://www.techinsider.io/airplane-flight-cosmic-radiation-exposure -altitude-2015-11.

"Chronic Pain," FamilyDoctor.Org, November 6, 2019, https://familydoctor.org/condition /chronic-pain/.

Mike Cummings, Asbjørn Hróbjartsson, Edzard Ernst, "Should Doctors Recommend Acupuncture for Pain?" *BMJ*, March 7, 2018, https://www.bmj.com/content/360/bmj.k970.

Joseph DeAcetis, "Lambs EMF Blocking Underwear: Protecting Men With Space-Grade Technology," *Forbes*, May 31, 2019, https://www.forbes.com/sites/josephdeacetis/2019/05/31 /lambs-is-protecting-men-with-space-grade-technology/#2909d2961b46.

Jancee Dunn, "More Celebrities Are Using Energy Healing—But Does It Work?" *Health*, September 19, 2019, https://www.health.com/mind-body/energy-healing.

Bruce Durie, "Senses Special: Doors of Perception," *New Scientist*, January 25, 2005, https:// www.newscientist.com/article/mg18524841-600-senses-special-doors-of-perception/.

Judy Estrin and Sam Gill, "The World Is Choking on Digital Pollution," *Washington Monthly*, March 2019, https://washingtonmonthly.com/magazine/january-february-march-2019/the -world-is-choking-on-digital-pollution/.

"Fundamentals of Oncology Acupuncture," Memorial Sloan Kettering Cancer Center, Continuing Education Program, https://www.mskcc.org/departments/survivorship-supportive-care /integrative-medicine/programs/fundamentals-oncology-acupuncture.

Ann Louise Gittleman, "The Health Effects of EMFs: How to Protect Yourself From the Dangers of Electromagnetic Radiation," *Conscious Lifestyle Magazine*, https://www.consciouslifestyle mag.com/emf-dangers-health-effects-radiation/.

Sophia Gushée, "Could 'Dirty Electricity' Be Affecting Your Health?" Well+Good/Good Home, March 19, 2018, https://www.wellandgood.com/good-home/dirty-electricity-from-light bulb -sophia-gushee/.

Isabel Pastor Guzman, "Tuning in to the Earth's Natural Rhythm," *Brain World Magazine*, October 4, 2017, https://brainworldmagazine.com/tuning-in-to-the-earths-natural-rhythm/.

Julien Happich, "What's Next after 5G? 6G at THz Frequencies," eeNews Europe, November 6, 2019, https://www.eenewseurope.com/news/whats-next-after-5g-6g-thz-frequencies.

Michael Hardy, "To These People, Electronic Devices Are the Enemy," *Wired*, February 4, 2020, https://www.wired.com/story/the-disputed-diagnosis-forcing-people-into-faraday-cages/.

Ned Herrmann, "What Is the Function of the Various Brainwaves?" *Scientific American*, December 22, 1997, https://www.scientificamerican.com/article/what-is-the-function-of-t-1997-12-22/.

Dominik Irnich, Johannes Fleckenstein, "Fascia: The Tensional Network of the Human Body," Science Direct, 2012, https://www.sciencedirect.com/topics/medicine-and-dentistry /acupuncture-point.

"Joseph Henry: A Life in Science, Electromagnetism," Smithsonian Archives, https://siarchives.si .edu/history/featured-topics/henry/electromagnetism.

Suresh Karve and Milind Bembalkar, "Life in the Grip of Electrosmog," Down to Earth, September 3, 2019, https://www.downtoearth.org.in/blog/india/life-in-the-grip-of-electrosmog -66483.

John Kelly, "How Nikola Tesla Worked," How Stuff Works, November 3, 2012, https://science .howstuffworks.com/innovation/famous-inventors/nikola-tesla3.htm.

Robert B. Kelly, "Acupuncture for Pain," American Academy of Family Physicians, September 1, 2009, https://www.aafp.org/afp/2009/0901/p481.html.

Jordan Kisner, "Reiki Can't Possibly Work. So Why Does It?" *Atlantic*, April 2020, https://

www.theatlantic.com/magazine/archive/2020/04/reiki-cant-possibly-work-so-why-does
-it/606808/.

Benjamin Kliger, Raymond Teets, Melissa Quick, "Complementary/Integrative Therapies That
Work: A Review of the Evidence," American Academy of Family Physicians, September 1,
2016, https://www.aafp.org/afp/2016/0901/p369.html.

Christy J. W. Ledford, Carla L. Fisher, David A. Moss, and Paul F. Crawford, "Critical Factors to
Practicing Medical Acupuncture in Family Medicine: Patient and Physician Perspectives,"
Journal of the American Board of Family Medicine 31, no. 2 (March 2018), https://www.jabfm
.org/content/31/2/236.long.

Tianjun Liu, "The Scientific Hypothesis of an 'Energy System' in the Human Body," Journal
of Traditional Chinese Medicine 5, no. 1 (January 2018), https://doi.org/10.1016/j
.jtcms.2018.02.003.

Sue Marek, "Marek's Take: Cyber-Physical Fusion, Satellites, and the 6G Promise," FierceWireless,
February 3, 2020, https://www.fiercewireless.com/5g/marek-s-take-cyber-physical-fusion
-satellites-and-6g-promise.

Marko Markov, "Expanding Use of Pulsed Electromagnetic Field Therapies," Journal of Electromagnetic Biology and Medicine 26, no. 3 (2007), https://doi.org/10.1080/15368370701
580806.

"Meridian (Chinese Medicine)," Wikipedia, December 22, 2019, https://en.wikipedia.org/wiki
/Meridian_(Chinese_medicine).

"Meridian System," Science Direct, https://www.sciencedirect.com/topics/medicine-and-dentistry
/meridian-system.

Richard Nahin, Robin Boineau, Partap Khalsa, Barbara Stussman, Wendy Weber, "Evidence-
Based Evaluation of Complementary Health Approaches for Pain Management in the United
States," Mayo Clinic Proceedings, September 2016, https://www.mayoclinicproceedings.org
/article/S0025-6196(16)30317-2/fulltext.

F. R. Nelson et al., "The Use of a Specific Pulsed Electromagnetic Field (PEMF) in Treating Early
Knee Osteoarthritis," Fifty-Sixth Annual Meeting of the Orthopaedic Research Society, New
Orleans, Louisiana, 2010, https://www.orthocormedical.com/wp-content/uploads/2016/01
/Nelson-2010-Henry-Ford-PEMF-Study.pdf.

Carl Pfaffmann, "Human Sensory Reception," Encyclopaedia Britannica, September 16, 1998,
https://www.britannica.com/science/human-sensory-reception.

Oska Phoenix, "The New Science of Human Energy Fields," The Way of Meditation, June 13,
2019, https://www.thewayofmeditation.com.au/the-new-science-of-human-energy-fields.

Arthur Pilla et al., "Electromagnetic Fields as First Messenger in Biological Signaling: Application
to Calmodulin-Dependent Signaling in Tissue Repair," Biochimica et Biophysica Acta,
December 2011, https://www.researchgate.net/publication/51725344_Electromagnetic
_fields_as_first_messenger_in_biological_signaling_Application_to_calmodulin-dependent
_signaling_in_tissue_repair.

Ross Pomeroy, "Why Is There Magnetite in the Human Brain?" RealClear Science, June 11,
2019, https://www.realclearscience.com/blog/2019/06/11/why_is_there_magnetite_in_the
_human_brain.html.

Kerry Redshaw, "Nikola Tesla (1856–1943)" KerryR.net, ABN 94 704 041 474 © 1996, Brisbane,
Queensland, Australia, http://www.kerryr.net/pioneers/tesla.htm.

Brianna Sacks, "Reiki Goes Mainstream: Spiritual Touch Practice Now Commonplace in Hospitals," Washington Post, May 16, 2014, https://www.washingtonpost.com/national/religion
/reiki-goes-mainstream-spiritual-touch-practice-now-commonplace-in-hospitals/2014/05/16
/9e92223a-dd37-11e3-a837-8835df6c12c4_story.html.

"Silver Nanowires Protect against Electromagnetic Radiation," The Engineer, October 28, 2019,
https://www.theengineer.co.uk/nanowires-electromagnetic-radiation/.

Erin Tallman, "DreamHomes in Malaysia: Building a Healthier Lifestyle," Medical Expo e-magazine,
October 2, 2019, http://emag.medicalexpo.com/dreamhomes-in-malaysia-building-a
-healthier-lifestyle/.

Jason Tanz, "Werner Herzog's Web," Wired, July 19, 2016, https://www.wired.com/2016/07
/warner-herzog-lo-and-behold/.

Maria Temming, "People Can Sense Earth's Magnetic Field, Brain Waves Suggest," Science News,
March 18, 2019, https://www.sciencenews.org/article/people-can-sense-earth-magnetic-field
-brain-waves-suggest.

"Transcranial Magnetic Stimulation," Mayo Clinic, https://www.mayoclinic.org/tests-procedures/transcranial-magnetic-stimulation/about/pac-20384625.

Walt Whitman, "I Sing the Body Electric," https://www.poetryfoundation.org/poems/45472/i-sing-the-body-electric.

Tracy V. Wilson, "How Wireless Power Works," How Stuff Works, January 12, 2007, https://electronics.howstuffworks.com/everyday-tech/wireless-power.htm.

02: Catching Nature's Vibe

American Geophysical Union, "Whale Song's Changing Pitch May Be Response to Population, Climate Changes," Phys Org, November 29, 2018, https://phys.org/news/2018-11-whale-songs-pitch-response-population.html.

"The Backster Effect: Are Plants Conscious?" Subtle Energy Science, https://subtle.energy/the-backster-effect-are-plants-conscious/.

Patrick Barkham, "Eva Meijer: 'Of Course Animals Speak. The Thing Is We Don't Listen,'" *Guardian*, November 13, 2019, https://www.theguardian.com/science/2019/nov/13/of-course-animals-speak-eva-meijer-on-how-to-communicate-with-our-fellow-beasts#img-1.

Christopher Bergland, "A New Da Vinci Code? Divine Proportion Found in Human Skulls," *Psychology Today*, October 4, 2019, https://www.psychologytoday.com/us/blog/the-athletes-way/201910/new-da-vinci-code-divine-proportion-found-in-human-skulls.

"'Bionic Mushrooms' Fuse Nanotech, Bacteria and Fungi," Stevens Institute of Technology, https://www.stevens.edu/news/bionic-mushrooms-fuse-nanotech-bacteria-and-fungi.

Teresa Carey, "Human Echolocators 'See' with Sound. Here's What That Actually Looks Like," PBS Science, September 8, 2017, https://www.pbs.org/newshour/science/human-echolocators-see-sound-heres-actually-looks-like.

Nick Carne, "Can We Beat Mosquitos at Their Own Game?" *Cosmos Magazine*, November 11, 2019, https://cosmosmagazine.com/biology/can-we-beat-mosquitos-at-their-own-game.

——, "Mushrooms Plus Bacteria Equals a New Source of Electrical Energy," *Cosmos Magazine*, November 8, 2018, https://cosmosmagazine.com/biology/mushrooms-plus-bacteria-equals-a-new-source-of-electrical-energy.

J. V. Chamary, "*Star Wars: The Last Jedi* Finally Explains the Force," *Forbes*, January 6, 2018, https://www.forbes.com/sites/jvchamary/2018/01/06/star-wars-last-jedi-force/#572e695b7a32.

Elephant Listening Project, "Deep into Infrasound," Cornell Lab, June 8, 2017, https://elephantlisteningproject.org/all-about-infrasound/.

Zach Fitzner, "Animals Use Infrasound to Communicate over Vast Distances," Earth.com, May 31, 2019, https://www.earth.com/news/animals-use-infrasound-communicate/.

Nic Fleming, "Plants Talk to Each Other Using an Internet of Fungus," BBC News, November 11, 2014, http://www.bbc.co.uk/earth/story/20141111-plants-have-a-hidden-internet.

Cal Flyn, "From Whirring Moths to Squeaking Bats, the World Is Full of Animal Communications We Cannot Detect," *Prospect Magazine*, June 8, 2019, https://www.prospectmagazine.co.uk/magazine/cal-flyn-whirring-moths-to-squeaking-bats-the-world-is-full-of-animal-communications-we-cannot-detect.

"The Force," Star Wars Databank, https://www.starwars.com/databank/the-force.

Robbie Gonzalez, "What It Takes to Be an Expert Human Echolocator," *Wired*, August 31, 2017, https://www.wired.com/2017/08/takes-expert-human-echolocator/.

Sophie Haigney, "The Lessons to Be Learned from Forcing Plants to Play Music," NPR, February 21, 2020, https://www.npr.org/2020/02/21/807821340/the-lessons-to-be-learned-from-forcing-plants-to-play-music.

Peter Heft, "Gilles Deleuze and Felix Guattari's *The Rhizome*, Translated from English to American," Mantle, June 26, 2017, https://www.themantle.com/philosophy/rhizome-american-translation.

Isochronic Entrainment Tones, "The Invisible Musicians—Whale Songs & Ocean Waves," Spotify, https://open.spotify.com/playlist/4ty0IwbMlfFKuhE6EpZ11X.

"Ley Lines: The Earth's Subtle Energy Grid," Subtle Energy Sciences, https://subtle.energy/ley-lines-the-earths-subtle-energy-grid/.

Clay B. Marsh, "There's a Hidden Language in Nature—a Golden Ratio of Healing," Thrive Global, April 23, 2019, https://thriveglobal.com/stories/spring-nature-and-the-golden-ratio/.

Ian Garrick Mason, "What Do We Really Mean by the 'Language' of Animals?" *Spectator*,

November 30, 2019, https://www.spectator.co.uk/2019/11/what-do-we-really-mean-by-the
-language-of-animals/.

Maryann Mott, "Did Animals Sense Tsunami Was Coming?" *National Geographic*, January 4,
2005, https://www.nationalgeographic.com/animals/2005/01/news-animals-tsunami-sense
-coming/.

David Newnham, "The Ley of the Land," *Guardian*, May 12, 2000, https://www.theguardian.com
/theguardian/2000/may/13/weekend7.weekend1.

Hannah Osborne, "Ancient Tree with Record of Earth's Magnetic Field Reversal in Its Rings
Discovered," *Newsweek*, July 4, 2019, https://www.newsweek.com/ancient-tree-discovered
-earths-magnetic-field-1447570.

Jennifer Ouellette, "Scientists Look to Music of the Volcanos to Better Monitor Eruptions," Ars
Technica, August 24, 2018, https://arstechnica.com/science/2018/08/scientists-look-to
-music-of-volcanos-to-better-monitor-eruptions/.

Katie Scott, "Oxford Scientists Discover Elephant 'Bee Warning' Call," *Wired*, April, 27, 2010,
https://www.wired.co.uk/article/oxford-scientists-discover-elephant-bee-call.

Iona McCombie Smith, "Could Plants Really Have Consciousness? The Scientific Experiment
That Reveals They Do," The BL, September 3, 2019, https://thebl.com/culture/could-plants
-really-have-consciousness-the-scientific-experiment-that-reveals-that-they.do.html.

Paul Stamets, "6 Ways Mushrooms Can Save the World," TED, March 2008, https://www.ted
.com/talks/paul_stamets_6_ways_mushrooms_can_save_the_world.

Shawn Stevenson, "The Benefits of Earthing and Grounding: How Touching the Earth Can
Improve Your Health," *Conscious Lifestyle Magazine*, April 25, 2019, https://www.conscious
lifestylemag.com/earthing-and-grounding-benefits/.

Carsten Welsch, "Star Wars: From the Force to R2-D2, Does the Science Hold Up?" Conversation, December 18, 2019, https://theconversation.com/star-wars-from-the-force-to-r2d2
-does-the-science-hold-up-128822.

"Whale Songs: Soundtrack of the Seas," Whale Watch Cabo, January 4, 2020, https://whale
watchcabo.com/whale-songs-soundtrack-of-the-seas.

Sam Wong, "Flowers Hear Bees and Make Sweeter Nectar When They're Buzzing Nearby," *New
Scientist*, January 8, 2019, https://www.newscientist.com/article/2189875-flowers-hear-bees
-and-make-sweeter-nectar-when-theyre-buzzing-nearby/.

Ed Yong, "Bacteria Unite to Form Living Electric Cables That Stretch for Centimetres," *Discover
Magazine*, October 24, 2012, https://www.discovermagazine.com/planet-earth/bacteria
-unite-to-form-living-electric-cables-that-stretch-for-centimetres.

——, "Plants Can Hear Animals Using Their Flowers," *Atlantic*, January 10, 2019, https://
www.theatlantic.com/science/archive/2019/01/plants-use-flowers-hear-buzz-animals/57
9964/.

Carl Zimmer, "Wired Bacteria Form Nature's Power Grid: 'We Have an Electric Planet,'" *New
York Times*, July 1, 2019, https://www.nytimes.com/2019/07/01/science/bacteria-microbes
-electricity.html.

03: The Body Light

Fraser A. Armstrong, "Photons in Biology," The Royal Society Publishing, October 6, 2013,
https://royalsocietypublishing.org/doi/10.1098/rsfs.2013.0039.

Samina T. Yousuf Azeemi and S. Mohsin Raza, "A Critical Analysis of Chromotherapy and Its
Scientific Evolution," *Evidence-Based Complementary and Alternative Medicine* 2, no.4 (December 2005): 481–88, https://www.ncbi.nlm.nih.gov/pmc/articles/PMC1297510/.

Bioregulatory Medicine, "Brmi: History—Dinshah Ghadiali," BRMI Bioregulatory Medicine Institute, https://www.brmi.online/dinshah-ghadiali.

Sharlene Breakey, "Does Light Therapy Work? Experts Explain Red, White, and Blue Light Therapies," Prevention.com, November 25, 2019, https://www.prevention.com/health/a29849201
/light-therapy/.

Fraser Cain, "How Long Does It Take Sunlight to Reach the Earth?" PhysOrg, April 15, 2013,
https://phys.org/news/2013-04-sunlight-earth.html.

"Chromotherapy Sauna Benefits: Color Therapy Explained," Sunlighten, March 4, 2020, https://
www.sunlighten.com/blog/chromotherapy-sauna-benefits-color-therapy-explained/.

"Continue Learning about LED Light Therapy," Holistic Light Therapy, https://www.holisticlight
therapy.co/led-light-therapy-resources/.

Tama Duffy Day, "The Healing Use of Color and Light," *Healthcare Design Magazine*, February 1, 2008, https://www.healthcaredesignmagazine.com/architecture/healing-use-light-and-color/.

Brian Dunbar, "What Is Color?" NASA.gov, https://www.nasa.gov/audience/forstudents/k-4/home/F_What_is_Color.html.

Jessica Estrada, "Your Breakdown of the Seven Unique Chakras' Colors and Meaning," Well+Good, November 28, 2019, https://www.wellandgood.com/good-advice/chakra-colors-and-meanings/.

Daniel Fels, "Cellular Communication through Light," *PLOS One*, April 1, 2009, https://journals.plos.org/plosone/article?id=10.1371/journal.pone.0005086.

Michelle Fondin, "What Is a Chakra?" Chopra Center, November 8, 2019, https://chopra.com/articles/what-is-a-chakra.

Holger Fuss, "The Struggle of Fritz-Albert Popp," Q-mag.org, https://q-mag.org/the-light-of-all-life.html.

Anne Helmenstine, "The Visible Spectrum: Wavelengths and Colors," ThoughtCo, April 2, 2020, https://www.thoughtco.com/understand-the-visible-spectrum-608329.

Asa Hershoff, "Science and the Rainbow Body, Part 1: Science Meets the Rainbow Body," BuddhistDoor.net, July 11, 2019, https://www.buddhistdoor.net/features/science-and-rainbow-bodies-part-1.

Kylie Holmes, "Sekhem—A Form of Ancient Egyptian Healing," PositiveHealth.com, April 2006, http://www.positivehealth.com/article/reiki/sekhem-a-form-of-ancient-egyptian-healing.

G. J. Hyland, "Herbert Fröhlich, FRS (1905–1991)–A Physicist Ahead of His Time," Taylor & Francis Online, July 9, 2009, https://www.tandfonline.com/doi/abs/10.1080/15368370500382164?scroll=top&needAccess=true&journalCode=iebm20&.

Institute for Quality and Efficiency in Health Care, "Does Light Therapy (Phototherapy) Help Reduce Psoriasis Symptoms?" National Center for Biotechnology Information, May 18, 2017, https://www.ncbi.nlm.nih.gov/books/NBK435696/.

Tega Jessa, "What Are Photons?" UniverseToday.com, September 19, 2010, https://www.universetoday.com/74027/what-are-photons/.

Tiina I. Karu, "Light Coherence: Is This Property Important for Photomedicine?" Institute of Laser and Information Technologies, Russian Academy of Sciences, June 4, 2011, http://photobiology.info/Coherence.html.

Won-Serk Kim and R. Glen Calderhead, "Is Light-Emitting Diode Phototherapy (LED-LLLT) Really Effective?" National Center for Biotechnology Information, 2011, https://www.ncbi.nlm.nih.gov/pmc/articles/PMC3799034/#__ffn_sectitle.

Debby Knox, "Latest Treatments for Depression," CBS4 Indianapolis, January 28, 2016, https://cbs4indy.com/news/4-your-health/latest-treatments-for-depression/.

Kara Ladd, "Is Chromotherapy the Real Deal?" *Architectural Digest*, March 20, 2019, https://www.architecturaldigest.com/story/is-chromotherapy-the-real-deal.

"Laser Therapy," Medlineplus Medical Encyclopedia, https://medlineplus.gov/ency/article/001913.htm.

"Let There Be Light," WRF.org, https://www.wrf.org/men-women-medicine/spectrochrome-dinshah-ghadiali.php.

Miriam Tomaz de Magalhães, Silvia Núñez, Ilka Kato, and Martha Ribeiro, "Light Therapy Modulates Serotonin Levels and Blood Flow in Women with Headache. A Preliminary Study," National Center for Biotechnology Information, January 23, 2016, https://www.ncbi.nlm.nih.gov/pubmed/26202374.

Kellee Maize, "Chakra Anatomy: How Chakral Function Relates to Human Physiology," Kellee Maize.com, https://www.kelleemaize.com/post/chakra-anatomy-how-chakral-function-relates-to-human-physiology.

MayoClinic, "Light Therapy," Mayo Foundation for Medical Education and Research, https://www.mayoclinic.org/tests-procedures/light-therapy/about/pac-20384604.

Sherri Melrose, "Seasonal Affective Disorder: An Overview of Assessment and Treatment Approaches," *Depression Research and Treatment* (November 25, 2015), https://pubmed.ncbi.nlm.nih.gov/26688752/.

W. S. Metcalf and T. I. Quickenden, "Mitogenetic Radiation," *Nature*, October 14, 1967, https://www.nature.com/articles/216169a0.

NASA/Marshall Space Flight Center, "NASA Space Technology Shines Light on Healing," *ScienceDaily*, December 21, 2000, www.sciencedaily.com/releases/2000/12/001219195848.htm.

Hugo J. Niggli, "Biophotons: Ultraweak Light Impulses Regulate Life Processes in Aging," *Journal of Gerontology & Geriatric Research* 3, no. 2 (2014): https://www.longdom.org/open-access/biophotons-ultraweak-light-impulses-regulate-life-processes-in-aging-2167-7182.1000143.pdf.

———, "Ultraweak Photons Emitted by Cells: Biophotons," *Journal of Photochemistry and Photobiology B: Biology* 14, nos. 1–2 (June 30, 1992): 144–46, https://doi.org/10.1016/1011-1344(92)85090-H.

Roni Caryn Rabin, "Blurred Vision, Burning Eyes: This Is a Lasik Success?" *New York Times*, June 11, 2018, https://www.nytimes.com/2018/06/11/well/lasik-complications-vision.html.

Emily Rekstis, "Healing Crystals 101," Healthline, June 21, 2018, https://www.healthline.com/health/mental-health/guide-to-healing-crystals#1.

Chandrasekhar Roychoudhuri, "The Nature of Light: What Are Photons," Spie.org, December 27, 2006, https://spie.org/news/0480-the-natureof-light-what-are-photons?SSO=1.

"Seasonal Affective Disorder," National Institute of Mental Health, https://www.nimh.nih.gov/health/topics/seasonal-affective-disorder/index.shtml.

Kendric C. Smith, "What Is Photobiology?" Stanford University School of Medicine, March 18, 2014, http://photobiology.info/introduction.html.

Ali Sundermier, "The Particle Physics of You," *Symmetry Magazine*, November 6, 2015, https://www.energy.gov/articles/particle-physics-you.

"10 Fascinating Facts about the Psychology of Color," Online Psychology Degree Guide, https://www.onlinepsychologydegree.info/psychology-color/.

Robin Turtle, "The Colours Associated with the Cardinal Directions in Chinese, Turkish, and Lakota," Polyglottando.com, August 11, 2016, https://www.polyglottando.com/?p=3444.

Christopher Wallis, "The Real Truth about the Chakras," UpliftConnect.com, February 13, 2016, https://upliftconnect.com/truth-about-the-chakras/.

Wellman Center for Photomedicine, "Why Light for Medicine?" Mass General Hospital, https://wellman.massgeneral.org/about-whylight.htm.

"What Are the 7 Chakras?" Yoga International, https://yogainternational.com/article/view/what-are-the-7-chakras.

"What Is the Origin of the Chakra System?" Indigo Massage and Wellness, March 26, 2013, https://indigomassagetherapy.com/uncategorized/what-is-the-origin-of-the-chakra-system/.

Magdalena Wszelaki, "The Healing Benefits of Infrared Saunas," HormonesBalance.com, November 6, 2019, https://hormonesbalance.com/articles/adrenals-articles/healing-benefits-infrared-saunas/.

04: Frequency Healing

Marisa Aveling, "The Doctor Is In (Your Pocket): How Apps Are Harnessing Music's Healing Powers," Pitchfork.com, November 14, 2016, https://pitchfork.com/features/article/9976-the-doctor-is-in-your-pocket-how-apps-are-harnessing-musics-healing-powers/.

Bend It Like Buddha, "The Science and Benefits of Chanting Mantras," Portugal Yoga Retreats, March 10, 2019, https://portugalyogaretreats.com/2019/03/10/the-science-and-benefits-of-chanting-mantras/.

Susan Blackmore, "Why Do Songs Get Stuck in My Head?" Science Focus, https://www.sciencefocus.com/the-human-body/why-do-songs-get-stuck-in-my-head/.

Charles Q. Choi, "These 3 Electroceuticals Could Help You Heal Faster," IEEE Spectrum, January 25, 2019, https://spectrum.ieee.org/semiconductors/devices/these-3-electroceuticals-could-help-you-heal-faster.

Jessica Kim Cohen, "Drug Companies Moving into Therapy Using Technology, Genetics," Modern Healthcare, August 24, 2019, https://www.modernhealthcare.com/care-delivery/drug-companies-moving-therapy-using-technology-genetics.

Zoe Cormier, "Music Therapy: The Power of Music for Health," Science Focus, April 15, 2020, https://www.sciencefocus.com/the-human-body/the-power-of-music-for-health/.

John Davis, "Can a Tuning Fork Detect a Stress Fracture?" Runners Connect, February 26, 2018, https://runnersconnect.net/tuning-fork-stress-fracture/.

Jennie Dear, "What It Feels Like to Die," *Atlantic*, September 21, 2016, https://www.theatlantic.com/health/archive/2016/09/what-it-feels-like-to-die/499319/.

Hannah Devlin, "Scientists Reverse Memory Decline Using Electrical Pulses," *Guardian*, April 8,

2019, https://www.theguardian.com/science/2019/apr/08/scientists-use-electrical-pulses-reverse-memory-decline-ageing.

"Egyptian Symbols: Sistrum," EgyptianGods.org, http://egyptian-gods.org/egyptian-symbols-sistrum/.

Kashmira Gander, "The Science Behind 'Killing' a Song When You Listen to It Too Much," *Independent*, May 10, 2017, https://www.independent.co.uk/life-style/killing-song-science-magic-lost-listen-too-much-sound-good-michael-bonshor-a7728156.html.

Alexandra Linnemann, Jana Strahler, and Urs M. Nater, "The Stress-Reducing Effect of Music Listening Varies Depending on the Social Context," *Psychoneuroendocrinology* 72 (June 23, 2016), https://pubmed.ncbi.nlm.nih.gov/27393906/.

Alexandra Linnemann, Mattes B. Kappert, Susanne Fischer, Johanna M. Doerr, Jana Strahler, and Urs M. Nater, "The Effects of Music Listening on Pain and Stress in the Daily Life of Patients with Fibromyalgia Syndrome," *Frontiers in Human Neuroscience* (2015), https://doi.org/10.3389/fnhum.2015.00434.

Meghan Neal, "The Many Colors of Sound," *Atlantic*, February 16, 2016, https://www.theatlantic.com/science/archive/2016/02/white-noise-sound-colors/462972/.

Kirsten Nunez, "What Is Pink Noise and How Does It Compare with Other Sonic Hues?" Healthline, June 21, 2019, https://www.healthline.com/health/pink-noise-sleep.

Gerald Oster, "Auditory Beats in the Brain," *Scientific American*, October 1973, https://www.scientificamerican.com/article/auditory-beats-in-the-brain/.

Alice Park, "Why It's Time to Take Electrified Medicine Seriously," *Time*, October 24, 2019, https://time.com/5709245/bioelectronic-medicine-treatments/.

Ernest K. J. Pauwels, "Mozart, Music, and Medicine," *Medical Principles and Practice: International Journal of the Kuwait University Health Science Centre* 23, no. 5 (2014): https://doi.org/10.1159/000364873.

Corinne Purtill, "Turns Out 'Sound Healing' Can Be Actually, Well, Healing," QZ.com, January 21, 2016, https://qz.com/595315/turns-out-sound-healing-can-be-actually-well-healing/.

"'Radiogenetics' Seeks to Remotely Control Cells and Genes," Rockefeller University, December 15, 2014, https://www.rockefeller.edu/news/9091-radiogenetics-seeks-to-remotely-control-cells-and-genes/.

Christian Reid, "The Science Behind Sound Healing," Sound Baths & Yoga with Christian, May 31, 2016, http://soundbathsandyoga.com/learn/2016/5/25/the-science-behind-sound-healing.

Casey Ross, "Once a Last Resort, This Pain Therapy Is Getting a New Life amid the Opioid Crisis," Stat News, January 23, 2019, https://www.statnews.com/2019/01/23/neuromodulation-pain-therapy-opioid-crisis/.

Rick Rowan, "A Brief History of Bioelectronic Medicine and Other Interesting Facts You Probably Didn't Know," NuroKor BioElectronics, January 9, 2019, https://www.nurokor.com/blog/history-of-bioelectronic-medicine.

Finn Saoirse, Daisy Fancourt, "The Biological Impact of Listening to Music in Clinical and Nonclinical Settings: A Systematic Review," *Progress in Brain Research* 237 (2018): 173–200, https://doi.org/10.1016/bs.pbr.2018.03.007.

Anne Trafton, "Ingestible Capsule Can Be Controlled Wirelessly," *MIT News*, December 13, 2018, http://news.mit.edu/2018/ingestible-pill-controlled-wirelessly-bluetooth-1213.

Lisa Trank, "Ancient Sound Technology: The Breath of Creation," Gaia.com, December 22, 2019, https://www.gaia.com/article/ancient-sound-technology-the-breath-of-creation.

05: Free Energy

Brendan Foster, "Einstein Killed The Aether. Now The Idea Is Back To Save Relativity," New Scientist, October 30, 2019 https://www.newscientist.com/article/mg24432543-300-einstein-killed-the-aether-now-the-idea-is-back-to-save-relativity/

Michael Greshko, "Cold Fusion Remains Elusive—But These Scientists May Revive the Quest," *National Geographic*, May 29, 2019, https://www.nationalgeographic.com/science/2019/05/cold-fusion-remains-elusive-these-scientists-may-revive-quest/.

Dr. Marc Halpern, "The Five Elements: Ether in Ayurveda," Ayurveda College, June 10, 2010, https://www.ayurvedacollege.com/blog/five-elements-ether-ayurveda/.

Ilker Koksal, "A Massive Investment: Google Announces 18 New Renewable Energy Deals," *Forbes*, October 12, 2019, https://www.forbes.com/sites/ilkerkoksal/2019/10/02/a-massive-investment-google-announces-18-new-renewable-energy-deals/#6602105c5024.

Neel V. Pateel, "WTF Is Zero Point Energy and How Could It Change the World?" Inverse, August 4, 2017, https://www.inverse.com/article/35077-wtf-is-zero-point-energy.

Helena Reilly, "Understanding the Power and Potential of Scalar Energy," Quantum Sound Therapy, June 8, 2017, https://quantumsoundtherapy.com/understanding-the-power-and -potential-of-scalar-energy/.

Jonathan Tennenbaum, "Cold Fusion: A Potential Energy Gamechanger," Asia Times, November 14, 2019, https://asiatimes.com/2019/11/cold-fusion-1-a-potential-energy-gamechanger/.

TV News Desk, "New Documentary TESLAFY ME Delves into Nikola Tesla," Broadway World, November 6, 2019, https://www.broadwayworld.com/bwwtv/article/New-Documentary -TESLAFY-ME-Delves-Into-Nikola-Tesla-20191106.

06: The Mind Field

Berit Brogaard, DMSci, PhD, "Brain Waves as Neural Correlates of Consciousness," Psychology Today, November 23, 2012, https://www.psychologytoday.com/us/blog/the-superhuman -mind/201211/brain-waves-neural-correlates-consciousness.

Brian Cox, "Deepak Chopra Doesn't Understand Quantum Physics, So Brian Cox Wants $1,000,000 from Him," New Statesman, July 2014, https://www.newstatesman.com/future-proof/2014/07 /deepak-chopra-doesnt-understand-quantum-physics-so-brian-cox-wants-1000000-him.

Carl Engelking, "Electrical Brain Stimulation Can Trigger Lucid Dreams," Discover Magazine, May 12, 2014, https://www.discovermagazine.com/mind/electrical-brain-stimulation-can-trigger -lucid-dreams.

Gaia Staff, "Akashic Records 101: Can We Access Our Akashic Records?" Gaia, September 17, 2019, https://www.gaia.com/article/akashic-records-101-can-we-access-our-akashic-records.

Jonathan Goldman, "Healing the Earth through Global Sound," Jonathan Goldman's Healing Sounds, https://www.healingsounds.com/healing-the-earth-through-global-sound/.

Tam Hunt, "The Hippies Were Right: It's All about Vibrations, Man!" Scientific American, December 5, 2018, https://blogs.scientificamerican.com/observations/the-hippies-were-right-its-all -about-vibrations-man/.

Mind Matters, "Panpsychism: You Are Conscious, But So Is Your Coffee Mug," Mind Matters News, November 13, 2018, https://mindmatters.ai/2018/11/panpsychism-you-are-conscious -but-so-is-your-coffee-mug/.

Ruby Mey, "10 Traits of Irresistible People with Good Vibes–Here Are Scientific Explanations," Epoch Times, March 28, 2019, https://www.theepochtimes.com/10-traits-of-irresistible -people-with-good-vibes-here-are-scientific-explanations_2821490.html.

Andrea Michelson, "Five Scientific Explanations for Spooky Sensations," Smithsonian Magazine, October 30, 2019, https://www.smithsonianmag.com/smart-news/five-scientific-explanations -spooky-sensations-180973436/.

News, "Why Some Scientists Believe the Universe Is Conscious," Mind Matters News, August 1, 2019, https://mindmatters.ai/2019/08/why-some-scientists-believe-the-universe-is-conscious/.

Valdas Noreika et al., "New Perspectives for the Study of Lucid Dreaming: From Brain Stimulation to Philosophical Theories of Self-Consciousness," International Journal of Dream Research 3, no. 1 (April 2010): 36–45, https://www.researchgate.net/publication/45360064 _New_perspectives_for_the_study_of_lucid_dreaming_From_brain_stimulation_to_philo sophical_theories_of_self-consciousness.

Meghan O'Gieblyn, "Do We Have Minds of Our Own?" New Yorker, December 4, 2019, https:// www.newyorker.com/books/under-review/do-we-have-minds-of-our-own.

Jessica D. Payne, "Sleep Makes Your Memories Stronger," Association for Psychological Science, November 12, 2010, https://www.psychologicalscience.org/news/releases/sleep-makes -your-memories-stronger.html.

"Psi Science," More Than Passing Strange, http://www.morethanpassingstrange.com/psi-science/.

Tibi Puiu, "Lucid Dreaming Easily Triggered by Zapping the Brain at 40Hz," ZME Science, May 12, 2014, https://www.zmescience.com/medicine/mind-and-brain/lucid-dreaming-electrical -trigger-042342/.

Amelia Tait, "Psychic Future: What's Next for the 'Precog Economy,'" Guardian, September 29, 2019, https://www.theguardian.com/global/2019/sep/29/psychic-future-what-next-for-the -precognition-economy.

"What Is Remote Viewing? Here's a Simple Explanation," Farsight Institute, https://farsight.org /WhatIsRemoteViewing.html.

Wired Staff, "Skeptic Revamps $1M Psychic Prize," *Wired*, January 12, 2007, https://www.wired.com/2007/01/skeptic-revamps-1m-psychic-prize/.

Kaitlyn Wylde, "What Are the Akashic Records?" Bustle, October 31, 2018, https://www.bustle.com/p/what-are-the-akashic-records-according-to-spirituality-every-soul-has-one-of-them-12966106.

Books

Avicenna, *Canon of Medicine*, ed. Laleh Bakhtiar, vol. 1 (Chicago: Kazi Publications, Inc., 1999).

Edwin D. Babbitt, *The Principles of Light and Color: The Classic Study of the Healing Power of Color* (New York, 1878).

Cleve Backster, *Primary Perception: Biocommunication with Plants, Living Foods, and Human Cells* (Anza, CA: White Rose Millennium Press, 2003).

Imants Baruss and Julia Mossbridge, *Transcendent Mind: Rethinking the Science of Consciousness* (Washington, DC: American Psychological Association, 2016).

David Chalmers, *The Character of Consciousness* (New York: Oxford University Press, 2010).

Theresa Cheung and Julia Mossbridge, *The Premonition Code: The Science of Precognition: How Sensing the Future Can Change Your Life* (New York: Watkins Publishing, 2018).

Constance Classen and David Howes, *Ways of Sensing,* (New York: Routledge, 2013).

Nick Cook, *The Hunt for Zero Point: Inside the Classified World of Antigravity Technology* (New York: Broadway Books, 2003).

Erik Davis, *High Weirdness: Drugs, Esoterica, and Visionary Experience in the Seventies* (Cambridge, MA: MIT Press, 2019).

Kingsley L. Dennis, *New Consciousness for a New World: How to Thrive in Transitional Times and Participate in the Coming Spiritual Renaissance* (Rochester, VT: Inner Traditions, 2011).

Dinshah Ghadiali, *Spectro-Chrome Metry Encyclopaedia* (Malaga, NJ: Spectro-Chrome Institute, 1939).

Ann Louise Gittleman, *Zapped: Why Your Cell Phone Shouldn't Be Your Alarm Clock and 1,268 Ways to Outsmart the Hazards of Electronic Pollution* (New York: HarperOne, 2010).

Bernard Haisch and Norman Dietz, *The God Theory: Universes, Zero-Point Fields, and What's Behind It All* (San Francisco: Weiser Books, 2006).

Nancy Humpel, *Journey to the Truth: An Introduction to the Reality of Ourselves and the World* (Calwell, Australia: Inspiring Publishers, 2013).

Moray B. King, *Tapping the Zero Point Energy* (Kempton, IL: Adventures Unlimited Press, 2002).

Hilma af Klint, *Notes and Methods*, ed. Christine Burgin (Chicago: University of Chicago Press, 2018).

Charles Klotsche, *Color Medicine: The Secrets of Color/Vibrational Healing* (Sedona, AZ: Light Technology Publishing, 1994).

Ervin László, *Science and the Akashic Field: An Integral Theory of Everything* (Rochester, VT: Inner Traditions, May 2007).

——, *The Akasha Paradigm: Revolution in Science, Evolution in Consciousness* (Cardiff-by-the-Sea, CA: Waterside Publications, 2012).

Bruce Lipton, *The Biology of Belief: Unleashing the Power of Consciousness, Matter & Miracles* (Carlsbad, CA: Hay House, Inc., 2016).

Lynne McTaggart, *The Field: The Quest for the Secret Force of the Universe* (New York: Harper Perennial, 2008).

Eva Meijer, *Animal Languages: The Secret Conversations of the Living World* (Cambridge, MA: MIT Press, 2019).

Vikas Nehru, *Global Wireless Spiderweb: The Invisible Threat Posed by Wireless Radiation* (Bloomington, IN: Partridge Publishing India, 2016).

Andrew Newberg, MD, and Mark Robert Waldman, *How Enlightenment Changes Your Brain: The New Science of Transformation* (New York: Avery, 2016).

Isaac Newton, *Philosophiæ Naturalis Principia Mathematica* (1687).

James L. Oschman, *Energy Medicine: The Scientific Basis* (London: Churchill Livingstone, 2015).

Katharine Payne, *Silent Thunder: In the Presence of Elephants* (New York: Penguin Books, 1999).

F. David Peat, *Synchronicity: The Bridge Between Matter and Mind* (New York: Bantam, 1987).

Jeremy Rifkin and David Cochran Heath, *The Green New Deal: Why the Fossil Fuel Civilization Will Collapse by 2028 and the Bold Economic Plan to Save Life on Earth* (New York: St. Martin's Press, 2019).

Jon Ronson, *The Men Who Stare at Goats* (New York: Simon & Schuster, 2009).

Oliver Sacks, *Musicophilia: Tales of Music and the Brain* (New York: Picador, 2018).

Marilyn Schlitz and Tina Amorok, with Mark S. Micozzi, *Consciousness and Healing: Integral Approaches to Mind-Body Medicine* (London: Churchill Livingstone, 2004).

Thomas D. Seeley, *Honeybee Democracy* (Princeton, NJ: Princeton University Press, 2010).

Rupert Sheldrake, *The Sense of Being Stared At: And Other Unexplained Powers of Human Minds* (South Paris, ME: Park Street Press, 2013).

Daniel Sullivan, *Ley Lines: The Greatest Landscape Mystery* (Glastonbury, UK: Green Magic Publishing, 2005).

Russell Targ, *The Reality of ESP: A Physicist's Proof of Psychic Abilities* (New York: Quest Books, 2012).

Carl Unger, *The Language of the Consciousness Soul: A Guide to Rudolf Steiner's "Leading Thoughts"*, ed. William Jens Jensen (Hudson, NY: SteinerBooks, 2012).

Thomas Valone, *Zero Point Energy: The Fuel of the Future* (Beltsville, MD: Integrity Research Institute, 2008).

Jaap Van Etten, *Gifts of Mother Earth: Earth Energies, Vortexes, Lines, and Grids* (Sedona, AZ: Light Technology Publishing, 2011).

Danah Zohar, *The Quantum Self* (New York: William Morrow, 1991).

Danah Zohar and Ian Marshall, *SQ: Connecting with Our Spiritual Intelligence* (New York: Bloomsbury USA, 2000).

Gary Zukav, *Dancing Wu Li Masters: An Overview of the New Physics* (New York: HarperOne, 2009).

People We Noted

Aboriginal Australians: https://en.wikipedia.org/wiki/Aboriginal_Australians.

Aristotle, Greek philosopher and polymath: en.wikipedia.org/wiki/Aristotle.

Michelle Addington, architect, engineer, Yale Graduate School of Design: https://soa.utexas.edu/people/michelle-addington.

Jhené Aiko, singer-songwriter: http://www.jheneaiko.com/.

Lawrence Anthony, the "Elephant Whisperer": https://en.wikipedia.org/wiki/Lawrence_Anthony.

Aaron Antonovsky, sociologist: https://jech.bmj.com/content/59/6/511.

Heidi Appel, dean of the Jesup Scott Honors College, University of Toledo: https://www.utoledo.edu/honors/dean.html.

Edwin D. Babbitt, physician, spiritualist, and early promoter of chromopathy: https://hymnary.org/person/Babbitt_Edwin.

Cleve Backster, former expert in lie detection with the CIA, creator of the theory of primary perception: https://en.wikipedia.org/wiki/Cleve_Backster.

Erykah Badu, singer-songwriter: http://www.erykah-badu.com/.

Emma Barratt and Nick Davis: https://peerj.com/blog/post/111369042754/emma-barratt-nick-davis-asmr/.

Dr. John Beaulieu, sound-healing pioneer, BioSonic Enterprises: https://biosonics.com/about-us/.

Robert O. Becker, orthopedic surgeon and researcher in electrophysiology/electromedicine: https://en.wikipedia.org/wiki/Robert_O._Becker.

Henri Bergson, philosopher: https://www.nobelprize.org/prizes/literature/1927/bergson/biographical/.

Robert Bigelow, founder and president of Bigelow Aerospace: bigelowaerospace.com.

Ray Bradbury, author and screenwriter: https://raybradbury.com/.

Anna Breytenbach, professional animal communicator trained by the Assisi International Animal Institute in California: https://www.animalspirit.org/.

Harold Saxton Burr, bioelectrics researcher: https://en.wikipedia.org/wiki/Harold_Saxton_Burr.

John Cage, composer: https://www.johncage.org/.

James Cameron, director: https://en.wikipedia.org/wiki/James_Cameron.

Bob Capranica, Cornell University acoustic biologist: https://digital.library.cornell.edu/catalog/ss574518.

David Carpenter, director of the University of Albany Institute for Health and the Environment: https://www.albany.edu/news/experts/8212.php.

Nicholas Christakis, sociologist and physician known for work on social networks: https://

en.wikipedia.org/wiki/Nicholas_Christakis; https://eeb.yale.edu/people/faculty-affiliated/nicholas-christakis.

Hans Christian Ørsted, physicist: https://en.wikipedia.org/wiki/Hans_Christian_%C3%98rsted.

Constance Classen, anthropologist: cmhp.ucsb.edu/team/schooler; http://centreforsensorystudies.org/members/constance-classen/.

Rex Cocroft, biologist, University of Missouri: https://cocroft.biology.missouri.edu/.

David Cohen, research scientist in biomagnetism, MIT: https://davidcohen.mit.edu/.

Jason Mark Cole, physicist, Imperial College London: http://jasmcole.github.io/.

Simone Dalla Bella, professor of psychology, Université de Montpellier: https://www.brams.org/en/membres/simone-dalla-bella/.

F. David Peat, theoretical and holistic physicist: fdavidpeat.com.

Erik Davis, author, podcaster, award-winning journalist, speaker: techgnosis.com/about/bio/.

Wade Davis, explorer-in-residence at National Geographic: https://www.nationalgeographic.org/find-explorers/e-wade-davis.

Gilles Deleuze, psychoanalyst and philosopher: https://plato.stanford.edu/entries/deleuze/.

René Descartes, philosopher, mathematician, and scientist: en.wikipedia.org/wiki/René_Descartes.

Émile Deschamps, poet: en.wikipedia.org/wiki/ Émile_Deschamps.

Billie Eilish, singer-songwriter: https://www.billieeilish.com/.

Michael Faraday, scientist: https://en.wikipedia.org/wiki/Michael_Faraday.

Sheldon M. Feldman, MD, surgeon, Montefiore Center for Cancer: https://www.montefiore.org/body.cfm?id=1735&action=detail&ref=9070.

Richard Feynman, theoretical physicist: richardfeynman.com.

Joe Firmage, chairman, Science Invents; founder and electrodynamic experimentalist, New PhysicsTeam.org: newphysicsteam.org/joseph-p-firmage.

Martin Fleischmann, electrochemist: en.wikipedia.org/wiki/Martin_Fleischmann.

James H. Fowler, social scientist: https://en.wikipedia.org/wiki/James_H._Fowler.

Herbert Fröhlich, physicist: https://en.wikipedia.org/wiki/Herbert_Fr%C3%B6hlich.

Luigi Galvani, professor of anatomy, University of Bologna: https://www.unibo.it/en/university/who-we-are/our-history/famous-people-guests-illustrious-students/luigi-galvani-1.

Michael Garstang, environmental scientist, University of Virginia: http://evsc.as.virginia.edu/people/profile/mxg.

Gabrielle Giffords, congresswoman: https://giffords.org/people/gabrielle-giffords/.

Jane Goodall, primatologist and anthropologist: https://www.janegoodall.org/.

Felix Guattari, psychoanalyst and philosopher: https://mitpress.mit.edu/contributors/felix-guattari.

Alexander Gurwitsch, biologist and medical scientist: https://www.biologicalmedicineinstitute.com/gurwitsch.

Lilach Hadany, mathematician, Tel Aviv University: https://www.hadanylab.com/lilach-hadany.

Bernard Haisch, astrophysicist: calphysics.org/haisch/.

Dr. Ernst Hartmann, medical doctor: https://en.wikipedia.org/wiki/Ernst_Hartmann.

Anne Marie Helmenstine, PhD, chemist at ThoughtCo: https://www.thoughtco.com/anne-marie-helmenstine-ph-d-601916.

Heinrich Hertz, physicist: https://en.wikipedia.org/wiki/Heinrich_Hertz.

Werner Herzog, filmmaker: https://wernerherzog.com/.

Albrecht Heyer, PhD, bionutritionist, Bionutrition Institute NY.

Mae-Wan Ho, geneticist, biophysicist: https://en.wikipedia.org/wiki/Mae-Wan_Ho.

Dr. Martin Hoffert, professor emeritus of physics, New York University: as.nyu.edu/content/nyu-as/as/departments/physics/people/faculty.html.

Tam Hunt, founder of Community Renewable Solutions LLC, scholar in philosophy and energy policy, lecturer at UC Santa Barbara's Bren School of Environmental Science & Management: communityrenewables.biz/about-us.html.

Motoji Ikeya, quantum geophysicist: http://www.davidjaybrown.com/dr-motoji-ikeya/.

Carl Jung, psychiatrist and psychoanalyst who founded analytical psychology: en.wikipedia.org/wiki/Carl_Jung.

Erin Kelly, All That's Interesting: https://allthatsinteresting.com/.

Raven Keyes, Reiki Master Teacher: https://www.ravenkeyesmedicalreiki.com/.

Joseph Kirschvink, geologist: https://en.wikipedia.org/wiki/Joseph_Kirschvink.

Daniel Kish, president of World Access for the Blind: https://visioneers.org/daniel-kish/.

Mitchell Krucoff, professor of medicine, cardiologist, Duke Clinical Research Institute: scholars
.duke.edu/person/Mitchell.Krucoff.

Ervin László, systems theorist: https://en.wikipedia.org/wiki/Ervin_L%C3%A1szl%C3%B3.

Helen Lavretsky, professor of psychiatry, UCLA: https://www.semel.ucla.edu/profile/helen
-lavretsky-md.

Jung-Ah Lee, PhD, RN: https://www.faculty.uci.edu/profile.cfm?faculty_id=5580.

Ed Leeper, physcist: http://www.emfservices.com/profile.htm.

Rabbi Leila Gal Berner, PhD: https://www.reconstructingjudaism.org/profile/rabbi-leila-gal-berner
-phd.

Michael Levin, PhD, biologist, Wyss Institute, Harvard University: https://wyss.harvard.edu/team
/associate-faculty/michael-levin-ph-d/.

Tianjun Liu, researcher, Beijing University of Chinese Medicine: http://english.bucm.edu.cn/.

Dr. Derek Lovley, microbiologist, University of Massachusetts at Amherst: https://www.micro
.umass.edu/faculty-and-research/derek-lovley.

Dr. Manu Mannoor, assistant professor of mechanical engineering, Stevens Institute of Technology: https://web.stevens.edu/facultyprofile/?id=2044.

Bob Marley, singer-songwriter: https://www.bobmarley.com/history/.

Rollin McCraty, psychophysiologist, director of research at HeartMath Institute: https://www
.heartmath.org/about-us/team/founder-and-executives/.

David E. McManus, PhD, research scientist: https://livingnow.com.au/scientists-journey-reiki/.

Ronald Melnick, PhD, NIH, retired: https://www.researchgate.net/profile/Ronald_Melnick.

Franz Mesmer, known for animal magnetism: https://en.wikipedia.org/wiki/Franz_Mesmer.

Dr. Edgar Mitchell, Apollo 14 astronaut, founder of the Institute of Noetic Sciences: noetic.org
/about/origins/.

J. P. Morgan, American financier: https://en.wikipedia.org/wiki/J._P._Morgan.

Peter Morgan, MD, PhD, chair of psychiatry, Lawrence and Memorial Hospital: medicine.yale
.edu/profile/peter_morgan/.

Meaghan Morrow, music therapist and certified brain injury specialist, TIRR Memorial Hermann
Rehabilitation Hospital: http://maeghanmorrow.com/about/.

Wolfgang Amadeus Mozart, composer: https://en.wikipedia.org/wiki/Wolfgang_Amadeus_Mozart.

Elon Musk, technology entrepreneur: https://en.wikipedia.org/wiki/Elon_Musk.

Richard L. Nahin, PhD, MPH, lead epidemiologist, NCCIH: https://www.nccih.nih.gov/about
/staff/richard-l-nahin.

Esra Neufeld, PhD, computational life sciences, IT'IS: https://itis.swiss/who-we-are/staff-members
/all-staff/esra-neufeld/.

Andrew Newberg, neuroscientist, professor and director of research, Marcus Institute of Integrative Health at Thomas Jefferson University and Hospital: andrewnewberg.com.

Isaac Newton, mathematician, physicist, astronomer, theologian, and author: en.wikipedia.org
/wiki/Isaac_Newton.

Lars Peter Nielsen, professor of microbial ecology, Aarhus University: https://pure.au.dk/portal
/en/lars.peter.nielsen@biology.au.dk.

Joe Patitucci, Jon Shapiro, and Alex Tyson, Data Garden: https://www.datagarden.org/who
-we-are.

Katy Payne, zoologist and bioacoustics biologist: https://www.penguinrandomhouse.com
/books/331558/silent-thunder-by-katy-payne/.

Michael Persinger, neuroscientist, professor of psychology and coordinator of the BSc in behavioral neuroscience, Laurentian University: laurentian.ca/faculty/arts/research-creativity.

Candace Pert, American neuroscientist: https://en.wikipedia.org/wiki/Candace_Pert.

Plato, philosopher: https://en.wikipedia.org/wiki/Plato.

Dr. Gerald Pollack, professor of bioengineering, University of Washington: pollacklab.org.

Stanley Pons, electrochemist: en.wikipedia.org/wiki/Stanley_Pons.

Fritz-Albert Popp, biophysicist: https://www.iumab.org/prof-fritz-albert-popp/.

Karl H. Pribram, neuroscientist: https://en.wikipedia.org/wiki/Karl_H._Pribram.

Hal Puthoff, PhD, theoretical physicist; founder of Earthtech International, Inc.; director of Institute for Advanced Studies: earthtech.org/team/.

Pythagoras, philosopher: https://www.ancient.eu/Pythagoras/.

Dean Radin, parapsychologist, chief scientist at the Institute of Noetic Sciences, Distinguished Professor at the California Institute of Integral Studies: deanradin.org.

James Randi, magician and escape artist, investigator and demystifier of paranormal and pseudoscientific claims: web.randi.org/about-james-randi.html.

Wilhelm Reich, doctor of medicine and psychoanalyst: https://www.wilhelmreichtrust.org /biography.html.

Baron Carl van Reichenbach, known for odic force: https://en.wikipedia.org/wiki/Carl _Reichenbach.

John Rogers, materials scientist, Northwestern University: https://www.farley.northwestern.edu /people/faculty/john-rogers.html.

Sigur Rós, rock band: https://sigurros.com/.

Alfonso Rueda, professor emeritus of electrical engineering, California State University: web .csulb.edu/divisions/aa/catalog/2012-2013/faculty/emeriti_n-r.html.

Dr. Oliver Sacks, former professor of neurology, Columbia University: https://en.wikipedia.org /wiki/Oliver_Sacks.

Dr. Teppo Särkämö, University of Helsinki: https://researchportal.helsinki.fi/en/persons/teppo -s%C3%A4rk%C3%A4m%C3%B6.

Marilyn Schlitz, PhD, social scientist, author, researcher, CEO/president emeritus and senior fellow at the Institute of Noetic Sciences: marilynschlitz.com.

Jonathan Schooler, psychologist, professor of psychological and brain sciences, director of Center for Mindfulness & Human Potential, University of California, Santa Barbara: psych .ucsb.edu/people/faculty/jonathan-schooler.

Erwin Schrödinger, founder of quantum theory: https://en.wikipedia.org/wiki/Erwin_Schr%C3 %B6dinger.

Winfried Otto Schumann, physicist: https://en.wikipedia.org/wiki/Winfried_Otto_Schumann.

Eric Schwitzgebel, philosopher, professor of philosophy, University of California, Riverside: faculty.ucr.edu/~eschwitz/.

Rupert Sheldrake, biologist and author best known for hypothesis of morphic resonance: sheldrake.org.

Susan Simard, ecologist, University of British Columbia: https://profiles.forestry.ubc.ca/person /suzanne-simard/.

George Soulié de Morant, scholar and diplomat: https://en.wikipedia.org/wiki/George_Souli %C3%A9_de_Morant.

Paul Stamets, mycologist: http://www.fungi.com/about-paul-stamets.html.

Rudolf Steiner, metaphysician, researcher, teacher, artist: rudolfsteiner.org.

Dr. John Stolz, microbiologist, Duquesne University: https://www.duq.edu/academics/faculty /john-stolz.

Henry Streby, ecologist, University of California, Berkeley: https://nature.berkeley.edu/beislab /BeissingerLab/?page_id=239.

Ingo Swann, parapsychologist, remote viewer, consciousness researcher, visionary artist: ingo swann.com.

Russell Targ, physicist, parapsychologist, author, cofounder of the Stanford Research Institute's investigation into psychic abilities in the 1970s and 1980s: espresearch.com.

Nikola Tesla, inventor and electrical engineer: https://en.wikipedia.org/wiki/Nikola_Tesla.

Lore Thaler, psychologist at Durham University: https://www.dur.ac.uk/psychology/staff/profile /?id=9604.

Dr. Jeffrey Thompson, founder/director of the Center for Neuroacoustic Research in Carlsbad, California: https://www.scientificsounds.com/.

Greta Thunberg, teenage environmental activist: facebook.com/gretathunbergsweden/.

Kevin Tracey, bioelectric medicine pioneer at the Feinstein Institutes for Medicine Research in New York: https://feinstein.northwell.edu/institutes-researchers/our-researchers/kevin-j -tracey-md.

Jacques Vallée, computer scientist, pioneer of the ARPANET, venture capitalist, author, scientific researcher of UFOs, astronomer: jacquesvallee.net.

Vitaly Vodyanoy, MS, PhD, professor of physiology, Auburn University: https://auburn.academia .edu/VitalyVodyanoy.

Ursula Voss, psychologist, professor, J.W. Goethe-University: user.uni-frankfurt.de/~voss/home page/en-engl/vita.html.

Xudong Wang, materials scientist, University of Wisconsin-Madison: https://energy.wisc.edu/about/energy-experts/xudong-wang.

Alfred Watkins, photographer and antiquarian, https://en.wikipedia.org/wiki/Alfred_Watkins

Weather Report, jazz fusion band: https://en.wikipedia.org/wiki/Weather_Report.

Rütger Wever, scientist: https://en.wikipedia.org/wiki/R%C3%BCtger_Wever.

John Archibald Wheeler, theoretical physicist, "Father of the Black Hole": https://phy.princeton.edu/department/history/faculty-history/john-wheeler.

Dr. Harry Whelan, professor of pediatric neurology, director of hyperbaric medicine, Medical College of Wisconsin: https://www.mcw.edu/find-a-doctor/whelan-harry-t-md.

Walt Whitman, poet: https://en.wikipedia.org/wiki/Walt_Whitman.

Heinrich Wilhelm Dove, https://en.wikipedia.org/wiki/Heinrich_Wilhelm_Dove.

Johann Wolfgang von Goethe, poet and scientist: https://en.wikipedia.org/wiki/Johann_Wolfgang_von_Goethe.

Nancy Wertheimer, epidemiologist: https://microwavenews.com/news-center/nancy-wertheimer-who-linked-magnetic-fields-childhood-leukemia-dies.

Yossi Yovel, ecologist, Tel Aviv University: https://english.tau.ac.il/profile/yossiy.

John Zimmerman, DC, chiropractor: https://www.johnzimmermandc.com/.

Danah Zohar, physicist, philosopher, author: danahzohar.com

Organizations

ActiPatch: www.ActiPatch.com.

Advanced Research Projects Agency (ARPA-E): www.arpa-e.energy.gov.

American Institutes for Research (AIR): www.air.org.

American Polarity Therapy Association: https://polaritytherapy.org/.

Animal-Computer Interaction Lab at the Open University (UK): http://www.open.ac.uk/blogs/ACI/.

Assisi International Animal Institute in California: https://www.assisianimals.org/.

BioSonic Enterprises: https://biosonics.com/.

Brigham and Women's Hospital: https://www.brighamandwomens.org/.

Buddhism and Science Research Lab, Centre of Buddhist Studies at the University of Hong Kong: https://www.buddhism.hku.hk/.

Cefaly Migraine Wearable Device: https://www.cefaly.us/en/migraine-treatment-cefaly.

Center for Reiki Research: www.centerforreikiresearch.org.

Children's Hospital of Wisconsin: https://chw.org/.

Clinic of Cognitive Neurology at the University Hospital of Leipzig, Germany: https://www.uniklinikum-leipzig.de/einrichtungen/tagesklinik-neurologie/en.

Clinical Psychology & Psychotherapy, University of Zürich, Switzerland: https://www.psychology.uzh.ch/en.html.

Columbia University Medical Center, Center for Comprehensive Wellness, Integrative Therapies: https://www.ccw.columbia.edu/patient-care/integrative-therapies.

Consciousness and Healing Initiative: https://www.chi.is/.

Cornell Lab's Elephant Listening Project: https://www.birds.cornell.edu/ccb/simple-list/elephant-listening-project/.

Data Garden: https://www.datagarden.org/who-we-are.

Disney's Animal Kingdom: https://disneyworld.disney.go.com/attractions/animal-kingdom/.

Dogs for the Disabled (UK): https://www.dogsforgood.org/.

Draper Laboratory: https://www.draper.com/.

Durham University: https://www.dur.ac.uk/.

Earthtech.org/The Institute for Advanced Studies at Austin: www.earthtech.org.

Farsight Institute: www.farsight.org.

Foundation for Research on Information Technologies in Society (IT'IS): https://itis.swiss/who-we-are/.

Framingham Heart Study: https://framinghamheartstudy.org/.

GlaxoSmithKline: https://www.gsk.com-en-gb/home/.

Global Coherence Initiative/HeartMath: www.heartmath.org/gci/.

Global Wellness Summit: https://www.globalwellnesssummit.com/.

Harvard Medical School Beth Israel Deaconess Medical Center: https://hms.harvard.edu/affiliates/beth-israel-deaconess-medical-center.

HeartMath Institute: www.heartmath.org.
Institute for Venture Science: www.ivscience.org.
Institute of Noetic Sciences (IONS): www.noetic.org.
Integratron: https://www.integratron.com/.
International EMF Scientist Appeal: https://emfscientist.org/.
ITER (International Nuclear Fusion Research): www.iter.org.
Johns Hopkins University School of Medicine: https://www.hopkinsmedicine.org/som/.
Journal of Geophysical Research: Oceans: https://agupubs.onlinelibrary.wiley.com/journal/21699291.
Marconi Union: https://marconiunion.com/.
Massachusetts General Hospital: https://www.massgeneral.org/.
Mayo Clinic: www.mayoclinic.org.
Medical College of Wisconsin in Milwaukee: https://www.mcw.edu/.
Medical Reiki Works: https://www.medicalreikiworks.org/.
MedRhythm: https://www.medrhythms.com/.
Memorial Sloan Kettering Integrative Medicine: https://www.mskcc.org/cancer-care/diagnosis-treatment/symptom-management/integrative-medicine.
MIT (Massachusetts Institute of Technology): http://www.mit.edu/.
Monroe Institute: www.monroeinstitute.org.
NASA: https://www.nasa.gov/.
National Center for Complementary and Integrative Health: https://www.nccih.nih.gov/.
National Council on Aging: https://www.ncoa.org/.
National Institute of Mental Health: https://www.nimh.nih.gov/index.shtml.
National Institutes of Health: https://www.nih.gov/.
New York State Department of Environmental Conservation: https://www.dec.ny.gov/.
New York University: https://www.nyu.edu/.
NextMind: www.next-mind.com.
NoiseZ: https://apps.apple.com/us/app/noisez-soothing-sleep-sounds/id662842897.
Northwestern University: https://www.northwestern.edu/.
Ohio State University Center for Regenerative Medicine and Cell-Based Therapies: https://medicine.osu.edu/departments/regenerative-medicine/cell-based-therapy-research.
Oska Wellness: www.oskawellness.com.
Oxford University: http://www.ox.ac.uk/.
Oxford University Sleep and Circadian Neuroscience Institute: https://www.ndcn.ox.ac.uk/research/sleep-circadian-neuroscience-institute.
Penn Medicine: https://www.pennmedicine.org/.
PLoS Computational Biology: https://journals.plos.org/ploscompbiol/.
Quell: https://www.quellrelief.com/.
Recovery Unplugged: https://www.recoveryunplugged.com/.
Rensselaer Polytechnic Institute: https://www.rpi.edu/.
Save the Elephants: https://www.savetheelephants.org/.
Science Invents: www.scienceinvents.org.
Soul Rider, LLC: www.thesoulrider.net.
Southall Environmental Associates: https://sea-inc.net/.
SQUID (Superconducting Quantum Interference Device): https://en.wikipedia.org/wiki/SQUID.
Stevens Institute of Technology: https://www.stevens.edu/.
Sync Project: http://syncproject.co/.
Tel Aviv University: https://english.tau.ac.il/.
Theranica: https://theranica.com/welcome/.
ThoughtCo: https://www.thoughtco.com/.
UC Davis Institute for Regenerative Cures: https://blog.cirm.ca.gov/tag/uc-davis-institute-for-regenerative-cures/.
United Nations Environment Programme: https://www.unenvironment.org/.
University of California, Berkeley: https://www.berkeley.edu/.
University of Illinois at Urbana-Champaign: https://illinois.edu/.
University of Missouri: https://www.umsystem.edu/.
University of Virginia: https://www.virginia.edu/.
University of Washington: https://www.washington.edu/.

Verily Life Sciences: https://verily.com/.

Washington Park Zoo, Portland, Oregon: https://www.oregonzoo.org/.

Way of Meditation: https://www.thewayofmeditation.com.au/.

WCS (Wildlife Conservation Society): https://www.wcs.org/.

Wellman Center for Photomedicine at Massachusetts General Hospital: https://wellman
.massgeneral.org/.

WiFi Angels: https://nextnature.net/projects/wifi-angels.

World Access for the Blind: https://waftb.net/.

World Health Organization, International Agency for Research on Cancer: https://www.who.int
/ionizing_radiation/research/iarc/en/.

Wyss Institute, Harvard University: https://wyss.harvard.edu/.

Yale Graduate School of Design: https://www.art.yale.edu/interactive-design.

Yellowstone National Park: https://www.nps.gov/yell/index.htm.

Yoga International: https://yogainternational.com/.

Acknowledgments

Alice in Futureland would not have been possible without the guidance of numerous individuals who in one way or another have contributed their valuable, innovative wisdom and energy.

First and foremost, we owe our deepest gratitude to our Sputnik Futures and Alice in Futureland colleagues and partners, Lisa, Amy, and Jordan, who have wandered with us through the many years of futures research. Know that we have circled the stars together and have many more laps to go. ☺

Thank you, Luis, for bringing Alice to life.

Special thanks to the like minds at Tiller Press, especially Theresa, who saw the vision for this future series; Sam for his insights; and Hannah for her sharp editing.

And to the many frontier thinkers who have shared their knowledge with enthusiasm, we are gratefully indebted.

To our families who have lived this journey with us, thank you all with love!

J&J

Notes

00: INTRODUCTION

1. Robert O. Becker and Gary Selden, *The Body Electric: Electromagnetism and the Foundation of Life* (New York: Quill William Morrow, 1985).
2. Lindsay Brownell, "Using Bioelectricity to Study How Cells Make Collective Decisions about Growth and Shape," Wyss Harvard News, July 26, 2019, https://wyss.harvard.edu/mike-levin-on-electrifying-insights-into-how-bodies-form/.
3. J. M. Herndon, Proceedings of the Royal Society of London, Series A, 368 (1979): 495; Journal of Geomagnetism and Geoelectricity 45 (1993): 423; Proceedings of the Royal Society of London, Series A, 445 (1994): 453; Proceedings of the National Academy of Sciences of the United States of America 93 (1996): 646; D. F. Hollenbach and J. M. Herndon, Proceedings of the National Academy of Sciences of the United States of America 98 (2001): 11085.
4. Wayne Hu, "Sound Out Origins," University of Chicago, http://background.uchicago.edu/~whu/SciAm/sym2.html.
5. Jeff Volk, "Sound Insights," MACROmedia 2011, http://www.cymaticsource.com/articles/a1-article.html.
6. Smooth Radio.com, July 26, 2018, https://www.smoothradio.com/features/beach-boys-best-songs-video/.
7. Ruby Mey, "The 10 Traits of Irresistible People with Good Vibes—Here Are Scientific Explanations," *Epoch Times*, March 3, 2019, https://www.theepochtimes.com/10-traits-of-irresistible-people-with-good-vibes-here-are-scientific-explanations_2821490.html.
8. Ibid.
9. Heidi Moawad, "How Real Are Vibes: The Good and the Bad?" Neurology Times, February 15, 2018, https://www.neurologytimes.com/blogs/how-real-are-vibes-good-and-bad.
10. HeartMath Institute, Science of the Heart: Exploring the Role of the Heart in Human Performance, chapter 6: Energetic Communication, https://www.heartmath.org/research/science-of-the-heart/energetic-communication/.

01: WE LIVE IN AN ELECTROMAGNETIC WORLD

1. Elizabeth Blaber, Kevin Sato, and Eduardo A. C. Almeida, "Stem Cell Health and Tissue Regeneration in Microgravity," *Stem Cells and Development* 23, supplement 1 (December 1, 2014): 73–78, doi:10.1089/scd.2014.0408.
2. Ju Hwan Kim, Jin-Koo Lee, Hyung-Gun Kim, Kyu-Bong Kim, and Hak Rim Kim, "Possible Effects of Radiofrequency Electromagnetic Field Exposure on Central Nerve System," *Biomolecules & Therapeutics* 27, no. 3 (May 2019): 265–75, doi:10.4062/biomolther.2018.152.
3. "Measuring Radiation," United States Nuclear Regulatory Commission, https://www.nrc.gov/about-nrc/radiation/health-effects/measuring-radiation.html.
4. "Heinrich Hertz," *Encyclopedia Britannica*, February 18, 2020, https://www.britannica.com/biography/Heinrich-Hertz.
5. John Kelly, "How Nikola Tesla Worked," HowStuffWorks.com, https://science.howstuffworks.com/innovation/famous-inventors/nikola-tesla.htm.
6. Gilbert King, "The Rise and Fall of Nikola Tesla and His Tower," *Smithsonian Magazine*, February 4, 2013, https://www.smithsonianmag.com/history/the-rise-and-fall-of-nikola-tesla-and-his-tower-11074324/.
7. Nikola Tesla quote, Goodreads, https://www.goodreads.com/quotes/tag/frequency.
8. "Energy: The Driver of Climate—Electromagnetic Radiation," Climate Science Investigations, NASA, http://www.ces.fau.edu/nasa/module-2/radiation-sun.php.
9. "Electromagnetism," Smithsonian Institute Archives, https://siarchives.si.edu/history/featured-topics/henry/electromagnetism.
10. "Electromagnetism: Historical Survey," *Encyclopedia Britannica*, https://www.britannica.com/science/electromagnetism/Historical-survey.
11. "The Science of Brainwaves," NeuroHealth.com, https://nhahealth.com/brainwaves-the-language/.

12. Isabel Pastor Guzman, "Tuning in to the Earth's Natural Rhythm," *Brain World Magazine*, October 4, 2017, https://brainworldmagazine.com/tuning-in-to-the-earths-natural-rhythm/.

13. R. Wever, "The Effects of Electric Fields on Circadian Rhythmicity in Men," *Life Sciences and Space Research* 8 (1970): 177–87, https://www.ncbi.nlm.nih.gov/pubmed/11826883.

14. Kevin S. Saroka and Michael A. Persinger, "Quantitative Evidence for Direct Effects between Earth-Ionosphere Schumann Resonances and Human Cerebral Cortical Activity," *International Letters of Chemistry, Physics and Astronomy* 39 (October 2014): 166–94.

15. Ned Herrmann, "What Is the Function of the Various Brainwaves?" *Scientific American*, December 22, 1997, https://www.scientificamerican.com/article/what-is-the-function-of-t-1997-12-22/.

16. Mayo Clinic Staff, "Transcranial Magnetic Stimulation," MayoClinic.org, https://www.mayoclinic.org/tests-procedures/transcranial-magnetic-stimulation/about/pac-20384625.

17. Beverly Rubik, PhD, David Muehsam, PhD, Richard Hammerschlag, PhD, and Shamini Jain, PhD, "Biofield Science and Healing: History, Terminology, and Concepts," *Global Advances in Health and Medicine* 4, supplement 1 (October 30, 2018): 8–14, https://doi.org/10.7453/gahmj.2015.038.suppl.

18. M. I. Faley, J. Dammers, Y. V. Maslennikov, J. F. Schneiderman, D. Winkler, V. P. Koshelets, N. J. Shah, and R. E. Dunin-Borkowski, "High-T_c SQUID Biomagnetometers," *Superconductor Science and Technology* 30 (2017): 083001, https://iopscience.iop.org/article/10.1088/1361-6668/aa73ad.

19. Marco Bischof, "The History of Bioelectromagnetism: The Instrument Era," in Mae-Wan Ho, Fritz-Albert Popp, and Ulrich Warnke, eds., *Bioelectrodynamics and Biocommunication* (Singapore: World Scientific Publishing Co., 1994).

20. Oska Phoenix, "The New Science of Human Energy Fields," Way of Meditation, June 13, 2019, https://www.thewayofmeditation.com.au/the-new-science-of-human-energy-fields.

21. Brownell, "Using Bioelectricity to Study How Cells Make Collective Decisions about Growth and Shape."

22. Michael Levin, "Bioelectric Mechanisms in Regeneration: Unique Aspects and Future Perspectives," *Seminars in Cell & Developmental Biology* 20, no. 5 (2009): 543–56, doi:10.1016/j.semcdb.2009.04.013.

23. Brownell, "Using Bioelectricity to Study How Cells Make Collective Decisions about Growth and Shape."

24. Amber Plante, "How the Human Body Uses Electricity," Graduate Student Association, University of Maryland Graduate School, https://www.graduate.umaryland.edu/gsa/gazette/February-2016/How-the-human-body-uses-electricity/.

25. Andy Fell, "Opposites Attract: How Cells and Cell Fragments Move in Electric Fields," UC Davis blog, March 28, 2013, http://blogs.ucdavis.edu/egghead/2015/10/12/the-sixth-sense-how-do-we-sense-electric-fields/.

26. Ananya Mandal, "Acupuncture History," New Medical Life Sciences, June 19, 2019, https://www.news-medical.net/health/Acupuncture-History.aspx.

27. F. Huang, *A-B Classic of Acupuncture and Moxibustion (Zhen Jiu Jia Yi Jing)* (Beijing: People's Health Publishers, 1956).

28. J. M. Helms, *Acupuncture Energetics: A Clinical Approach for Physicians* (Berkeley, CA: Medical Acupuncture Publishers, 1995).

29. Ding-zhong Li, Song-tao Fu, and Xiu-zhang Li, "Relationship of Meridians with Human Material and Energy Systems: Study on Theory and Clinical Application of Meridians (IV)," *Chinese Acupuncture & Moxibustion* 25, no. 2 (February 2005): 115–18.

30. Elizabeth Palermo, "What Is Qigong?" LiveScience, March 10, 2015, https://www.livescience.com/38192-qigong.html.

31. George Dillman with Chris Thomas, *Advanced Pressure Point Fighting of Ryukyu Kempo: Dillman Theory for All Systems* (Dillman Karate International, 1994).

32. Oska Phoenix, "The New Science of Human Energy Fields," *The Way of Meditation Blog*, June 13, 2019, https://www.thewayofmeditation.com.au/the-new-science-of-human-energy-fields.

33. Deborah Bleecker, *Acupuncture Points Handbook: A Patient's Guide to the Locations and Functions of Over 400 Acupuncture Points* (Draycott Publishing, 2017).

34. Tianjun Liu, "The Scientific Hypothesis of an 'Energy System' in the Human Body," *Journal of Traditional Chinese Medical Sciences* 5, no. 1 (January 2018): 29–34, https://doi.org/10.1016/j.jtcms.2018.02.003.

35. Hector W. H. Tsang, William W. N. Tsang, Alice Y. M. Jones, Kelvin M. T. Fung, Alan H. L. Chan, Edward P. Chan, and Doreen W. H. Au, "Psycho-Physical and Neurophysiological Effects of Qigong on Depressed Elders with Chronic Illness," *Aging & Mental Health* 17, no. 3 (2013): 336–48.

36. Stacey Nemour, "Qigong: Unleash Incredible Healing Powers," *HuffPost*, July 21, 2010, https://www.huffpost.com/entry/qigong-unleash-incredible_b_651561.

37. Walter Johnson, Oyere Onuma, Mayowa Owolabi, and Sonal Sachdev, "Stroke: A Global Response Is Needed," *Bulletin of the World Health Organization* 94 (2016): 634–34A, doi: http://dx.doi.org/10.2471/BLT.16.181636.

38. David E. McManus, "Reiki Is Better Than Placebo and Has Broad Potential as a Complementary Health Therapy," *Journal of Evidence-Based Complementary & Alternative Medicine* 22, no. 4 (October 2017): 1051–57, doi:10.1177/2156587217728644.

39. Institute of Medicine Committee on Advancing Pain Research and Education, *Relieving Pain in America: A Blueprint for Transforming Prevention, Care, Education, and Research* (Washington, DC: National Academies Press, 2011).

40. "Fundamentals of Oncology Acupuncture," Memorial Sloan Kettering Cancer Center, https://www.mskcc.org/departments/survivorship-supportive-care/integrative-medicine/programs/fundamentals-oncology-acupuncture.

41. Charles Martin, "Auburn Scientist Discovers Microstructure of Primo-Vascular System, Revealing Possible Foundation of How Acupuncture Works," Newsroom, Auburn University, December 5, 2016, http://ocm.auburn.edu/newsroom/news_articles/2016/12/auburn-scientist-discovers-microstructure-of-primo-vascular-system.php.

42. Richard L. Nahin, Robin Boineau, Partap S. Khalsa, Barbara J. Stussman, and Wendy J. Weber, "Evidence-Based Evaluation of Complementary Health Approaches for Pain Management in the United States," *Mayo Clinic Proceedings* 91, no. 9 (September 2016): 1292–306, doi:10.1016/j.mayocp.2016.06.007.

43. Jordan Kisner, "Reiki Can't Possibly Work. So Why Does It?" *Atlantic*, April 2020, https://www.theatlantic.com/magazine/archive/2020/04/reiki-cant-possibly-work-so-why-does-it/606808/.

44. Fangfang Ma, Xun Li, Yichen Wang, Ning Liang, Shixia Pan, Guoyan Yang, Yan Liao, Cong Zhang, Qingyi Zhang, and Yin Lin, "Effectiveness of Traditional Chinese Exercises on Stroke Risk Factors in Individuals with Pre-Hypertension or Mild-to-Moderate Essential Hypertension: A Systematic Review and Meta-Analysis," *Journal of Traditional Chinese Medical Sciences* 5, no. 3 (July 2018): 222–36, https://doi.org/10.1016/j.jtcms.2018.09.002.

45. Medical Reiki Works, https://www.medicalreikiworks.org/.

46. Center for Comprehensive Wellness, Columbia University Irving Medical Center.

47. Natalie L. Dyer, Ann L. Baldwin, and William L. Rand, "A Large-Scale Effectiveness Trial of Reiki for Physical and Psychological Health," *Journal of Alternative and Complementary Medicine* 25, no. 12 (December 2019): 1156–62, doi:10.1089/acm.2019.0022.

48. Francis H. Vala, MD, *The Third Vision: The Science of Personal Transformation* (Bloomington, IN: Balboa Press, 2012), 150.

49. Yan Zhang, Lixing Lao, Haiyan Chen, and Rodrigo Ceballos, "Acupuncture Use among American Adults: What Can Acupuncture Practitioners Learn from National Health Interview Survey 2007?" *Evidence-Based Complementary and Alternative Medicine* 2012 (2012): 710750, doi:10.1155/2012/710750.

50. James Dahlhamer, PhD, Jacqueline Lucas, MPH, Carla Zelaya, PhD, Richard Nahin, PhD, Sean Mackey, MD, PhD, Lynn DeBar, PhD, Robert Kerns, PhD, Michael Von Korff, ScD, Linda Porter, PhD, and Charles Helmick, MD, "Prevalence of Chronic Pain and High-Impact Chronic Pain

among Adults—United States, 2016," Centers for Disease Control and Prevention, September 14, 2018, https://www.cdc.gov/mmwr/volumes/67/wr/mm6736a2.htm; Richard L. Nahin, Bryan Sayer, Barbara J. Stussman, and Termeh M. Feinberg, "Eighteen-Year Trends in the Prevalence of and Health Care Use for Non-Cancer Pain in the United States: Data from the Medical Expenditure Panel Survey," *Journal of Pain* 20, no. 7 (July 1, 2019): 796–809, doi:10.1016/j.jpain.2019.01.003.

51. "Economic Toll of Opioid Crisis in U.S. Exceeded $1 Trillion Since 2001," Altarum, February 13, 2018, https://altarum.org/news/economic-toll-opioid-crisis-us-exceeded-1-trillion-2001; "Mental Disorders," World Health Organization, November 28, 2019, https://www.who.int/news room/fact-sheets/detail/mental-disorders.

52. National Academies of Sciences, Engineering, and Medicine; Health and Medicine Division; Board on Global Health; Board on Health Sciences Policy; Global Forum on Innovation in Health Professional Education; Forum on Neuroscience and Nervous System Disorders; Clare Stroud, Sheena M. Posey Norris, and Lisa Bain, eds., *The Role of Nonpharmacological Approaches to Pain Management: Proceedings of a Workshop* (Washington, DC: National Academies Press, 2019), https://www.ncbi.nlm.nih.gov/books/NBK541702/.

53. James L. Oschman, *Energy Medicine: The Scientific Basis* (Edinburgh: Elsevier, 2015), 250.

54. McCraty, Bradley, and Tomasino, 2004, heartmath.org.

55. HeartMath Institute, *Science of the Heart: Exploring the Role of the Heart in Human Performance*, chapter 1: Heart-Brain Communication, https://www.heartmath.org/research/science-of-the-heart/heart-brain-communication/.

56. James L. Oschman, "Science Measures the Human Energy Field," International Center for Reiki Training, https://www.Reiki.org/articles/science-measures-human-energy-field.

57. G. Telegdy, "The Action of ANP, BNP and Related Peptides on Motivated Behavior in Rats," *Reviews in the Neurosciences* 5, no. 4 (October–December 1994): 309–15.

58. J. Gutkowska, M. Jankowski, S. Mukaddam-Daher, and S. M. McCann, "Oxytocin Is a Cardiovascular Hormone," *Brazilian Journal of Medical and Biological Research* 33, no. 6 (June 2000): 625–33.

59. Franz Halberg et al., "Cross-Spectrally Coherent ~10.5- and 21-Year Biological and Physical Cycles, Magnetic Storms and Myocardial Infarctions," *Neuroendocrinology Letters* 21, no. 3 (2000): 233–58.

60. Karl H. Pribram, *Brain and Perception: Holonomy and Structure in Figural Processing* (Hillsdale, NJ: Lawrence Erlbaum Associates, 1991).

61. Karl H. Pribram and F. T. Melges, "Psychophysiological Basis of Emotion," in P. J. Vinken and G. W. Bruyn, eds., *Handbook of Clinical Neurology* (Amsterdam: North-Holland Publishing Company, 1969), 316–41.

62. HeartMath Institute, *Science of the Heart*, chapter 6: Energetic Communication.

63. Candace Pert, *Molecules of Emotion: The Science Behind Body-Mind Medicine* (New York: Simon & Schuster, 1999).

64. Rahasya Poe, *To Believe or Not to Believe: The Social and Neurological Consequences of Belief Systems* (Xlibris Corporation, 2009).

65. Nicholas A. Christakis and James H. Fowler, *Connected: The Surprising Power of Our Social Networks and How They Shape Our Lives* (New York: Little, Brown Spark, 2011).

66. James H. Fowler and Nicholas A. Christakis, "Dynamic Spread of Happiness in a Large Social Network: Longitudinal Analysis over 20 Years in the Framingham Heart Study," *BMJ* 337 (2008): a2338.

67. Robert O. Becker, *Cross Currents: The Perils of Electropollution, the Promise of Electromedicine* (New York: TarcherPerigee, 1990).

68. Suresh Karve and Milind Bembalkar, "Life in the Grip of Electrosmog," *Down to Earth*, September 3, 2019, https://www.downtoearth.org.in/blog/pollution/life-in-the-grip-of-electro smog-66483.

69. Elfide Gizem Kivrak, Kiymet Kübra Yurt, Arife Ahsen Kaplan, Isinsu Alkan, and Gamze

Altun, "Effects of Electromagnetic Fields Exposure on the Antioxidant Defense System," *Journal of Microscopy and Ultrastructure* 5, no. 4 (October–December 2017): 167-76, doi:10.1016/j.jmau.2017.07.003.

70. Kim et al., "Possible Effects of Radiofrequency Electromagnetic Field Exposure on Central Nerve System."

71. Chelsea E. Langer et al., "Patterns of Cellular Phone Use among Young People in 12 Countries: Implications for RF Exposure," *Environment International* 107 (October 2017): 65-74.

72. Crina Boros, "Mobile Phones and Health: Is 5G Being Rolled Out Too Fast?" *ComputerWeekly*, April 24, 2019, https://www.computerweekly.com/feature/Mobile-phones-and-health-is-5G-being-rolled-out-too-fast.

73. "Electromagnetic Fields and Public Health: Electromagnetic Hypersensitivity," World Health Organization, December 2005, https://www.who.int/peh-emf/publications/facts/fs296/en/.

74. Igor Belyaev et al., "EUROPAEM EMF Guideline 2016 for the Prevention, Diagnosis and Treatment of EMF-Related Health Problems and Illnesses," *Reviews on Environmental Health* 31, no. 3 (July 25, 2016): 363-97, doi:10.1515/reveh-2016-0011.

75. Jason Tanz, "Werner Herzog's Web," *Wired*, July 19, 2016, https://www.wired.com/2016/07/warner-herzog-lo-and-behold/.

76. L. Lloyd Morgan, Santosh Kesari, and Devra Lee Davis, "Why Children Absorb More Microwave Radiation Than Adults: The Consequences," *Journal of Microscopy and Ultrastructure* 2, no. 4 (December 2014): 197-204, https://doi.org/10.1016/j.jmau.2014.06.005.

77. David L. Means and Kwok W. Chan, "Evaluating Compliance with FCC Guidelines for Human Exposure to Radiofrequency Electromagnetic Fields," supplement C (edition 97-01) to OET bulletin 65 (edition 97-01), Federal Communications Commission Office of Engineering & Technology, June 2001, 16.

78. Morgan, Kesari, and Davis, "Why Children Absorb More Microwave Radiation Than Adults."

79. Joseph DeAcetis, "Lambs Is Protecting Men with Space-Grade Technology," *Forbes*, May 31, 2019, https://www.forbes.com/sites/josephdeacetis/2019/05/31/lambs-is-protecting-men-with-space-grade-technology/#1d5aa1b61b46.

80. Tribute to Robert O. Becker, https://codepen.io/tchuang/full/GZxJmz.

81. "U.N. Environment Programme Urged to Protect Nature and Humankind from Electromagnetic Fields (EMF)," Business Wire, July 22, 2019, https://www.businesswire.com/news/home/20190722005154/en/U.N.-Environment-Programme-Urged-Protect-Nature-Humankind.

82. Nancy Wertheimer and Ed Leeper, "Electrical Wiring Configurations and Childhood Cancer," *American Journal of Epidemiology* 109, no. 3 (March 1979): 273-84, 10.1093/oxfordjournals.aje.a112681.

83. Maria Feychting, William T. Kaune, David A. Savitz, and Anders Ahlbom, "Estimating Exposure in Studies of Residential Magnetic Fields and Cancer: Importance of Short-Term Variability, Time Interval between Diagnosis and Measurement, and Distance to Power Line," *Epidemiology* 7, no. 3 (May 1996): 220-24, www.jstor.org/stable/3702853.

84. Michael Segell, "Is 'Electrosmog' Harming Our Health?" NBCNews.com, January 18, 2010, http://www.nbcnews.com/id/34509513/ns/health-cancer/t/electrosmog-harming-our-health/#.XtO2JxNKigR.

85. Judy Estrin and Sam Gill, "The World Is Choking on Digital Pollution," *Washington Monthly*, January/February/March 2019, https://washingtonmonthly.com/magazine/january-february-march-2019/the-world-is-choking-on-digital-pollution/.

86. Michael J. Aminoff, François Boller, and Dick F. Swaab, "We Spend about One-Third of Our Life Either Sleeping or Attempting to Do So," *Handbook of Clinical Neurology* 98 (2011): vii, doi:10.1016/B978-0-444-52006-7.00047-2.

87. Igor Gorpinchenko, Oleg Nikitin, Oleg Banyra, and Alexander Shulyak, "The Influence of Direct Mobile Phone Radiation on Sperm Quality," *Central European Journal of Urology* 67, no. 1 (2014): 65-71, doi:10.5173/ceju.2014.01.art14.

88. James L. Oschman, "Perspective: Assume a Spherical Cow: The Role of Free or Mobile Electrons in Bodywork, Energetic and Movement Therapies," *Journal of Bodywork and Movement Therapies* 12, no. 1 (January 2008): 40–57, doi: 10.1016/j.jbmt.2007.08.002.

89. Connie X. Wang et al., "Transduction of the Geomagnetic Field as Evidenced from Alpha-Band Activity in the Human Brain," *eNeuro* 6, no. 2 (March 18, 2019), doi:10.1523/ENEURO.0483-18.2019.

90. Thomas J. Goodwin, PhD, "Physiological and Molecular Genetic Effects of Time-Varying Electromagetic Fields on Human Neuronal Cells," NASA, September 2003, https://ntrs.nasa.gov/archive/nasa/casi.ntrs.nasa.gov/20030075722.pdf.

91. Goodwin et al., United States Patent 8,795,147 B1, https://ntrs.nasa.gov/archive/nasa/casi.ntrs.nasa.gov/20150003336.pdf.

92. "Bio Osteogen System 204," FDA Premarket Approval P790002 S002, August 20, 1982, https://fda.report/PMA/P790002S002.

93. Klaus Martiny, Marianne Lunde, and Per Bech, "Transcranial Low Voltage Pulsed Electromagnetic Fields in Patients with Treatment-Resistant Depression," *Biological Psychiatry* 68, no. 2 (July 15, 2010): 163–69, doi:10.1016/j.biopsych.2010.02.017. PMID 20385376.

94. Carl Pfaffmann, "Human Sensory Reception," *Encyclopedia Britannica*, https://www.britannica.com/science/human-sensory-reception#ref531278.

95. Wang et al. "Transduction of the Geomagnetic Field as Evidenced from Alpha-Band Activity in the Human Brain."

96. Ibid.

97. Maria Temming, "People Can Sense Earth's Magnetic Field, Brain Waves Suggest," *Science News*, March 18, 2019, https://www.sciencenews.org/article/people-can-sense-earth-magnetic-field-brain-waves-suggest.

02: CATCHING NATURE'S VIBE

1. Diane Toomey, "Exploring How and Why Trees 'Talk' to Each Other," Yale Environment 360, September 1, 2016, https://e360.yale.edu/features/exploring_how_and_why_trees_talk_to_each_other.

2. Ibid.

3. Hannah Osborne, "Ancient Tree with Record of Earth's Magnetic Field Reversal in Its Rings Discovered," *Newsweek*, July 4, 2019, https://www.newsweek.com/ancient-tree-discovered-earths-magnetic-field-1447570.

4. Cheol Seong Jang, Terry L. Kamps, D. Neil Skinner, Stefan R. Schulze, William K. Vencill, and Andrew H. Paterson, "Functional Classification, Genomic Organization, Putatively cis-Acting Regulatory Elements, and Relationship to Quantitative Trait Loci of Sorghum Genes with Rhizome-Enriched Expression," *Plant Physiology* 142, no. 3 (November 2006): 1148–59, doi:10.1104/pp.106.082891.PMC1630734.PMID16998090.

5. Gilles Deleuze and Felix Guattari, *A Thousand Plateaus: Capitalism and Schizophrenia* (Minneapolis: University of Minnesota Press, 1987).

6. E. A. Solomon et al., "Widespread Theta Synchrony and High-Frequency Desynchronization Underlies Enhanced Cognition," *Nature Communications* 8 (November 2017), article 1704, doi:10.1038/s41467-017-01763-2.

7. "'Perfect Pitch' in Humans Far More Prevalent than Expected," University of Rochester, August 25, 2008, https://www.rochester.edu/news/show.php?id=3236.

8. Tom Huizenga, "How Do You Bond With Mozart? Adopt a Starling," NPR, April 20, 2017, https://www.npr.org/sections/deceptivecadence/2017/04/20/524349771/how-do-you-bond-with-mozart-adopt-a-starling.

9. Carl Zimmer, "Wired Bacteria Form Nature's Power Grid: We Have an Electric Planet," *New York Times*, July 1, 2019.

10. "Magnetite," Wikipedia, https://en.wikipedia.org/wiki/Magnetite.

11. J. L. Kirschvink, A. Kobayashi-Kirschvink, and B. J. Woodford, "Magnetite Biomineralization in the Human Brain," *Proceedings of the National Academy of Sciences of the United States of America* 89, no. 16 (August 15, 1992): 7683–87, doi:10.1073/pnas.89.16.7683.

12. Derek R. Lovley et al., "Geobacter: The Microbe Electric's Physiology, Ecology, and Practical Applications," *Advances in Microbial Physiology* 59 (2011): 1–100, doi:10.1016/B978-0-12-387661-4.00004-5.

13. Nils Risgaard-Petersen et al., "Cable Bacteria in Freshwater Sediments," *Applied and Environmental Microbiology* 81, no. 17 (September 2015): 6003–11, doi:10.1128/AEM.01064-15.

14. Abdelrhman Mohamed, Phuc T. Ha, Brent M. Peyton, Rebecca Mueller, Michelle Meagher, and Haluk Beyenal, "*In situ* Enrichment of Microbial Communities on Polarized Electrodes Deployed in Alkaline Hot Springs," *Journal of Power Sources* 414 (February 2019): 547–56, doi:10.1016/j.jpowsour.2019.01.027.

15. Naira Quintana, Frank Van der Kooy, Miranda D. Van de Rhee, Gerben P. Voshol, and Robert Verpoorte, "Renewable Energy from Cyanobacteria: Energy Production Optimization by Metabolic Pathway Engineering," *Applied Microbiology and Biotechnology* 91 (June 2011): 471–90, doi:10.1007/s00253-011-3394-0.

16. "Biomass Explained," US Energy Information Administration, June 21, 2018, https://www.eia.gov/energyexplained/biomass/.

17. Nick Carne, "Mushrooms Plus Bacteria Equals a New Source of Electrical Energy," *Cosmos Magazine*, November 6, 2018, https://cosmosmagazine.com/biology/mushrooms-plus-bacteria-equals-a-new-source-of-electrical-energy.

18. Sudeep Joshi, Ellexis Cook, and Manu S. Mannoor, "Bacterial Nanobionics via 3D Printing," *Nano Letters* 18, no. 12 (November 7, 2018): 7448–56, https://doi.org/10.1021/acs.nanolett.8b02642.

19. Nic Fleming, "Plants Talk to Each Other Using an Internet of Fungus," BBC.com, November 11, 2014, http://www.bbc.com/earth/story/20141111-plants-have-a-hidden-internet.

20. Pace, Matthew, "Hidden Partners: Mycorrhizal Fungi and Plants," New York Botanical Garden, https://sciweb.nybg.org/science2/hcol/mycorrhizae.asp.html.

21. *Avatar* Quotes, IMDB, 2009, https://www.imdb.com/title/tt0499549/quotes/qt1131653.

22. Paul Stamets, "Earth's Natural Internet," *Whole Earth*, fall 1999.

23. Ronald Hutton, *The Pagan Religions of the Ancient British Isles: Their Nature and Legacy* (Oxford, UK: Blackwell, 1991).

24. Brian Regal, "Ley Lines," in *Pseudoscience: A Critical Encyclopedia* (Westport, CT: Greenwood, 2009), 103.

25. David Newnham, "The Ley of the Land," *Guardian*, May 12, 2000, https://www.theguardian.com/theguardian/2000/may/13/weekend7.weekend1.

26. "Earth's Rays Part 3: What Is the Hartmann Grid?" Swiss Harmony, https://swissharmony.com/earth-rays/what-is-the-hartmann-grid/.

27. Ibid.

28. Cyndi Dale, *The Subtle Body Practice Manual: A Comprehensive Guide to Energy Healing* (Louisville, CO: Sounds True, 2013).

29. Newnham, "The Ley of the Land."

30. Charles R. Kelley, PhD, *What Is Orgone Energy?* (Vancouver, WA: K/R Publications, 1962).

31. Jeffrey B. Johnson, Richard C. Aster, and Philip R. Kyle, "Volcanic Eruptions Observed with Infrasound," *Geophysical Research Letters* 31, no. 14 (July 2004), doi:10.1029/2004GL020020.

32. Katy Payne, *Silent Thunder: In the Presence of Elephants* (New York: Penguin, 1999).

33. Motoji Ikeya, *Earthquakes and Animals: From Folk Legends to Science* (Singapore: World Scientific Publishing Co. Pte. Ltd., 2004).

34. Peggy S. M. Hill, *Vibrational Communication in Animals* (Cambridge, MA: Harvard University Press, 2008).

35. Reginald B. Cocroft and Rafael L. Rodríguez, "The Behavioral Ecology of Insect Vibrational Communication," *BioScience* 55, no. 4 (April 2005): 323–34, doi:10.1641/0006-3568(2005)055[0323:TBEOIV]2.0.CO;2.

36. Princeton University, Woodrow Wilson School of Public and International Affairs, "Forest

Soundscapes Monitor Conservation Efforts Inexpensively, Effectively," *ScienceDaily*, January 3, 2019, http://www.sciencedaily.com/releases/2019/01/190103185537.htm.

37. Ed Yong, "Plants Can Hear Animals Using Their Flowers," *Atlantic*, January 10, 2019.

38. Marine Veits et al., "Flowers Respond to Pollinator Sound within Minutes by Increasing Nectar Sugar Concentration," *Ecology Letters* 22, no. 9 (September 2019): 1483–92, https://doi .org/10.1111/ele.13331.

39. E. Wassim Chehab, Elizabeth Eich, and Janet Braam, "Thigmomorphogenesis: A Complex Plant Response to Mechano-Stimulation," *Journal of Experimental Botany* 60, no. 1 (January 2009): 43–56, doi:10.1093/jxb/ern315.

40. Jeff Sossamon, "Plants Respond to Leaf Vibrations Caused by Insects' Chewing, MU Study Finds," University of Missouri News Bureau, July 1, 2014, https://munewsarchives .missouri.edu/news-releases/2014/0701-plants-respond-to-leaf-vibrations-caused-by -insects%E2%80%99-chewing-mu-study-finds/.

41. Cleve Backster, *Primary Perception: Biocommunication with Plants, Living Foods, and Human Cells* (White Rose Millennium Press, 2003).

42. Ibid.

43. Music of the Plants, https://www.musicoftheplants.com/.

44. PlantWave, Data Garden, www.plantwave.com.

45. Jane E. Brody, "SCIENTIST AT WORK: Katy Payne: Picking Up Mammals' Deep Notes," *New York Times*, November 9, 1993.

46. Emmanuelle C. Leroy, Jean-Yves Royer, Julien Bonnel, and Flore Samaran, "Long-Term and Seasonal Changes of Large Whale Call Frequency in the Southern Indian Ocean," *Journal of Geophysical Research: Oceans* 123, no. 11 (November 2018): 8568–80, doi:10.1029/2018JC014352.

47. Wildlife Conservation Society, "Scientists Listen to Whales, Walruses and Seals in a Changing Arctic Seascape," PhysOrg, February 3, 2020, https://phys.org/news/2020-02 -scientists-whales-walruses-arctic-seascape.html.

48. M. Brock Fenton and John M. Ratcliffe, "Sensory Biology: Echolocation from Click to Call, Mouth to Wing," *Current Biology* 24, no. 24 (December 15, 2014): R1160–R1162, https://www .sciencedirect.com/science/article/pii/S0960982214014213.

49. Lore Thaler et al., "Mouth-Clicks Used by Blind Expert Human Echolocators–Signal Description and Model-Based Signal Synthesis," *PLoS Computational Biology* 13, no. 8 (August 31, 2017): e1005670, https://doi.org/10.1371/journal.pcbi.1005670.

50. Zach Fitzner, "Many Animals Use Infrasound to Communicate over Vast Distances," Earth .com, May 31, 2019, https://www.earth.com/news/animals-use-infrasound-communicate/.

51. Katharine B. Payne, William R. Langbauer Jr., and Elizabeth M. Thomas, "Infrasonic Calls of the Asian Elephant (*Elephas maximus*)," *Behavioral Ecology and Sociobiology* 18, no. 4 (1986): 297–301.

52. William R. Langbauer Jr., Katharine B. Payne, Russell A. Charif, Lisa Rapaport, and Ferrel Osborn, "African Elephants Respond to Distant Playbacks of Low-Frequency Conspecific Calls," *Journal of Experimental Biology* 157 (1991): 35–46.

53. Michael Garstang, Robert E. Davis, Keith Leggett, Oliver W. Frauenfeld, Steven Greco, Edward Zipser, and Michael Peterson, "Response of African Elephants (*Loxodonta africana*) to Seasonal Changes in Rainfall," *PloS One* 9, no. 10 (October 2014): e108736, https://doi .org/10.1371/journal.pone.0108736.

54. Cal Flyn, "From Whirring Moths to Squeaking Bats, the World Is Full of Animal Communications We Cannot Detect," *Prospect Magazine*, June 8, 2019, https://www.prospectmagazine .co.uk/magazine/cal-flyn-whirring-moths-to-squeaking-bats-the-world-is-full-of-animal-commu nications-we-cannot-detect.

55. Lucy E. King, Joseph Soltis, Iain Douglas-Hamilton, Anne Savage, and Fritz Vollrath, "Bee Threat Elicits Alarm Call in African Elephants," *PloS One* (April 26, 2010), https://doi .org/10.1371/journal.pone.0010346.

56. Karen McComb, David Reby, Lucy Baker, Cynthia Moss, and Soila Sayialel, "Long-

Distance Communication of Acoustic Cues to Social Identity in African Elephants," *Animal Behaviour* 65, no. 2 (February 2003): 317–29, https://www.sciencedirect.com/science/article/pii/S0003347203920471?via%3Dihub.

57. Jenna Benson, "Elephants Walk 12 Hours to Mourn Man Who Rescued Them," WFSM 101.7, October 2, 2019, https://www.wsfm.com.au/newsroom/elephants-walk-12-hours-to-mourn-man-who-rescued-them/.

58. Elephant Listening Project, "Deep into Infrasound," Cornell Lab, June 9, 2017, https://elephantlisteningproject.org/all-about-infrasound/.

59. Charles O. Choi, "Japan Quake May Have Struck Atmosphere First," LiveScience.com, October 10, 2011, https://www.livescience.com/16471-japan-quake-struck-atmosphere.html.

60. Henry M. Streby, Gunnar R. Kramer, Sean M. Peterson, Justin A. Lehman, David A. Buehler, and David E. Andersen, "Tornadic Storm Avoidance Behavior in Breeding Songbirds," *Current Biology* 25, no. 1 (January 2015): 98–102, https://doi.org/10.1016/j.cub.2014.10.079.

61. David Roos, "The 2004 Tsunami Wiped Away Towns with 'Mind-Boggling' Destruction," History.com, October 2, 2018, https://www.history.com/news/deadliest-tsunami-2004-indian-ocean.

62. Maryann Mott, "Did Animals Sense Tsunami Was Coming?" *National Geographic*, January 4, 2005, https://www.nationalgeographic.com/animals/2005/01/news-animals-tsunami-sense-coming/.

63. Ibid.

64. Mick Hamer, "Tigers Use Infrasound to Warn Off Rivals," *New Scientist*, May 2, 2003, https://www.newscientist.com/article/dn3680-tigers-use-infrasound-to-warn-off-rivals/.

65. Ian Garrick-Mason, "What Do We Really Mean by the 'Language' of Animals?" *Spectator*, November 30, 2019, https://www.spectator.co.uk/2019/11/what-do-we-really-mean-by-the-language-of-animals/.

66. "The Language of Bees," *Perfect Bee* (blog), February 11, 2016, https://www.perfectbee.com/blog/the-language-of-bees.

67. Nick Carne, "Can We Beat Mosquitos at Their Own Game?" *Cosmos Magazine*, November 10, 2019, https://cosmosmagazine.com/biology/can-we-beat-mosquitos-at-their-own-game.

68. Sid Perkins, "What's the Buzz: A New Mosquito Lure," Science News for Students, May 13, 2015, https://www.sciencenewsforstudents.org/article/whats-buzz-new-mosquito-lure.

69. Jason Daley, "Scientists Use AI to Decode the Ultrasonic Language of Rodents," *Smithsonian Magazine*, January 16, 2019, https://www.smithsonianmag.com/smart-news/scientists-use-ai-to-decode-ultrasonic-rodent-language-180971286/.

70. Helena Pozniak, "Call of the Wild: The New Science of Human-Animal Communication," *Guardian*, January 27, 2020, https://www.theguardian.com/global/2020/jan/27/call-of-the-wild-the-new-science-of-human-animal-communication.

71. Jane Goodall Quotes, Goodreads.com, https://www.goodreads.com/quotes/396319-the-least-i-can-do-is-speak-out-for-those.

72. "Katy Payne, In the Presence of Elephants and Whales," *On Being with Krista Tippett*, American Public Media, February 1, 2007, https://onbeing.org/programs/katy-payne-in-the-presence-of-elephants-and-whales/#transcript.

73. Michael Theys, "How to Communicate with Animals with Anna Breytenbach," AfricaFreak, February 4, 2020, https://africafreak.com/anna-breytenbach.

74. David Nabhan, "Science Hasn't Seen It All on Animal Perception," Newsmax, August 6, 2019, https://www.newsmax.com/davidnabhan/china-earthquakes-laquila-bees/2019/08/06/id/927569/.

75. Arri Eisen and Gary Laderman, eds., *Science, Religion, and Society: An Encyclopedia of History, Culture, and Controversy* (New York: Routledge, 2015).

76. Natalie Wolchover, "Did Leonardo da Vinci Copy His Famous 'Vitruvian Man'?" Live Science, January 30, 2012, https://www.livescience.com/18183-leonardo-da-vinci-copy-famous-vitruvian-man.html.

77. Rafael J. Tamargo, MD, and Jonathan A. Pindrik, MD, "Mammalian Skull Dimensions and the Golden Ratio (Φ)," *Journal of Craniofacial Surgery* 30, no. 6 (September 2019): 1750–55, doi:10.1097/SCS.0000000000005610.

78. Kara Rogers, "Biophilia Hypothesis," *Encyclopedia Britannica*, https://www.britannica.com/science/biophilia-hypothesis.

79. Hannah Ritchie and Max Roser, "Urbanization," OurWorldInData.org, November 2019, https://ourworldindata.org/urbanization.

80. US Environmental Protection Agency, "Report to Congress on Indoor Air Quality," vol. 2, EPA/400/1-89/001C, 1989, Washington, DC.

81. Gaétan Chevalier, Stephen T. Sinatra, James L. Oschman, and Richard M. Delany, "Earthing (Grounding) the Human Body Reduces Blood Viscosity—A Major Factor in Cardiovascular Disease," *Journal of Alternative and Complementary Medicine* 19, no. 2 (February 2013): 102–10, doi:10.1089/acm.2011.0820.

82. James L. Oschman, Gaétan Chevalier, and Richard Brown, "The Effects of Grounding (Earthing) on Inflammation, the Immune Response, Wound Healing, and Prevention and Treatment of Chronic Inflammatory and Autoimmune Diseases," *Journal of Inflammation Research* 8 (2015): 83–96, doi:10.2147/JIR.S69656.

83. Sylvia Thompson, "Take a 'Biophilia Tour' at the National Botanic Gardens. You'll Feel Better," *Irish Times*, January 11, 2020, https://www.irishtimes.com/life-and-style/health-family/take-a-biophilia-tour-at-the-national-botanic-gardens-you-ll-feel-better-1.4134889.

84. "Immerse Yourself in a Forest for Better Health," New York State Department of Environmental Conservation, https://www.dec.ny.gov/lands/90720.html.

03: THE BODY LIGHT

1. "The Science of HeartMath," HeartMath, https://www.heartmath.com/science/.

2. B. Lindström and M. Eriksson, "Professor Aaron Antonovsky (1923–1994): The Father of the Salutogenesis," *Journal of Epidemiology & Community Health* 59, no. 6 (June 2005): 511.

3. Khawaja Shamsuddin Azeemi, *Colour Therapy* (Karachi: Al-Kitab Publications, 1999).

4. David H. Frisch and Alan M. Thorndike, *Elementary Particles* (Princeton, NJ: David Van Nostrand Company, 1964), 22.

5. Emrys W. Evans, Charlotte A. Dodson, Kiminori Maeda, Till Biskup, C. J. Wedge, and Christiane R. Timmel, "Magnetic Field Effects in Flavoproteins and Related Systems," *Interface Focus* 3, no. 5 (October 2013), doi:10.1098/rsfs.2013.0037.

6. Katja A. Lamia et al., "AMPK Regulates the Circadian Clock by Cryptochrome Phosphorylation and Degradation," *Science* 326, no. 5951 (October 2009): 437–40, doi:10.1126/science.1172156.

7. Fraser A. Armstrong, "Photons in Biology," *Interface Focus* 3, no. 5 (October 2013): 20130039, doi:10.1098/rsfs.2013.0039.

8. Sharbel Weidner Maluf, Wilner Martínez-López, and Juliana da Silva, "DNA Damage: Health and Longevity," *Oxidative Medicine and Cellular Longevity* 2018 (September 2018): 9701647, doi:10.1155/2018/9701647.

9. W. Nagl and F. A. Popp, "A Physical (Electromagnetic) Model of Differentiation. 1. Basic Considerations," *Cytobios* 37, no. 145 (1983): 45–62.

10. Jiin-Ju Chang, Joachim Fisch, and Fritz-Albert Popp, eds., *Biophotons* (Boston: Kluwer Academic Publishers, 1998).

11. Hugo J. Niggli, "Ultraweak Photons Emitted by Cells: Biophotons," *Journal of Photochemistry and Photobiology B: Biology* 14, nos. 1–2 (June 1992): 144–46, doi:10.1016/1011-1344(92)85090-h.

12. Nagl and Popp, "A Physical (Electromagnetic) Model of Differentiation."

13. "Seasonal Affective Disorder," National Institute of Mental Health, https://www.nimh.nih.gov/health/topics/seasonal-affective-disorder/index.shtml.

14. Ibid.

15. Ibid.

16. "Why Light for Medicine?" Wellman Center for Photomedicine, https://wellman.mass general.org/about-whylight.htm.

17. Asa Hershoff, "Science and the Rainbow Body, Part 1: Science Meets the Rainbow Body," Buddhist Door, July 11, 2019, https://www.buddhistdoor.net/features/science-and-rainbow -bodies-part-1.

18. Patricia Pozo-Rosich, "Chronic Migraine: Its Epidemiology and Impact," *Revista de Neurologia* 54, supplement 2 (April 2012): S3–S11.

19. Ariovaldo Alberto da Silva Junior et al., "Temporomandibular Disorders and Chronic Daily Headaches in the Community and in Specialty Care," *Headache* 53, no. 8 (September 2013): 1350–55.

20. Miriam Tomaz de Magalhães, Silvia Cristina Núñez, Ilka Tiemy Kato, and Martha Simões Ribeiro, "Light Therapy Modulates Serotonin Levels and Blood Flow in Women with Headache. A Preliminary Study," *Experimental Biology and Medicine (Maywood, NJ)* 241, no. 1 (January 2016): 40–45, doi:10.1177/1535370215596383.

21. Dr. Adrian Larsen, "Laser vs. LED: What's the Difference?" Acupuncture Technology News, https://www.miridiatech.com/news/2014/02/laser-vs-led-whats-the-difference/.

22. "Light Therapy," Mayo Clinic, https://www.mayoclinic.org/tests-procedures/light-therapy /about/pac-20384604.

23. "Laser Therapy," MedlinePlus, US National Library of Medicine, https://medlineplus.gov /ency/article/001913.htm.

24. "Does Light Therapy (Phototherapy) Help Reduce Psoriasis Symptoms?" InformedHealth .org, Institute for Quality and Efficiency in Health Care (IQWiG), May 18, 2017, https://www.ncbi .nlm.nih.gov/books/NBK435696/.

25. NASA/Marshall Space Flight Center, "NASA Space Technology Shines Light on Healing," ScienceDaily, December 21, 2000, www.sciencedaily.com/releases/2000/12/001219195848 .htm.

26. Tama Duffy Day, "The Healing Use of Light and Color," *Healthcare Design Magazine*, February 1, 2008, https://www.healthcaredesignmagazine.com/architecture/healing-use-light -and-color/.

27. Samina T. Yousuf Azeemi and S. Mohsin Raza, "A Critical Analysis of Chromotherapy and Its Scientific Evolution," *Evidence-Based Complementary and Alternative Medicine* 2, no. 4 (December 2005): 481–88, doi:10.1093/ecam/neh137.

28. "What Is Color?" Student Features, NASA, March 11, 2004, https://www.nasa.gov/audience /forstudents/k-4/home/F_What_is_Color.html.

29. Anne Marie Helmenstine, PhD, "The Visible Spectrum: Wavelengths and Colors," ThoughtCo, April 2, 2020, https://www.thoughtco.com/understand-the-visible-spectrum -608329.

30. Azeemi and Raza, "A Critical Analysis of Chromotherapy and Its Scientific Evolution."

31. Ibid.

32. Ibid.

33. Charles Klotsche, *Color Medicine: The Secrets of Color/Vibrational Healing* (Sedona, AZ: Light Technology Publishing, 1993).

34. "Blue Light Has a Dark Side," Harvard Health Publishing, August 13, 2018, https://www .health.harvard.edu/staying-healthy/blue-light-has-a-dark-side.

35. University of Oxford, "Lighting Color Affects Sleep, Wakefulness," ScienceDaily, June 8, 2016, www.sciencedaily.com/releases/2016/06/160608154233.htm.

36. "Chromotherapy Sauna Benefits: Color Therapy Explained," Sunlighten.com, https:// www.sunlighten.com/blog/chromotherapy-sauna-benefits-color-therapy-explained/.

37. Azeemi and Raza, "A Critical Analysis of Chromotherapy and Its Scientific Evolution."

38. "Chromotherapy Sauna Benefits."

39. Ibid.

40. Ibid.

41. Rodrigo Noseda et al., "Migraine Photophobia Originating in Cone-Driven Retinal Pathways," *Brain* 139, no. 7 (July 2016): 1971–86, https://doi.org/10.1093/brain/aww119.

42. "Chromotherapy Sauna Benefits."

43. Ibid.

44. Ibid.

45. "What Are the Seven Chakras?" Yoga International, https://yogainternational.com/article/view/what-are-the-7-chakras.

46. Dinshah P. Ghadiali, *Spectro-Chrome Metry Encyclopedia* (Vineland, NJ: Dinshah Health Society, 1997).

47. Jessica Estrada, "Your Breakdown of the 7 Unique Chakra Colors and Meanings," Well+Good, November 28, 2019, https://www.wellandgood.com/good-advice/chakra-colors-and-meanings/.

48. Fraser Cain, "How Long Does It Take Sunlight to Reach the Earth?" PhysOrg, April 15, 2013, https://phys.org/news/2013-04-sunlight-earth.html.

04: FREQUENCY HEALING

1. Klotsche, *Color Medicine*.

2. Mark C. Genovese et al., "First-in-Human Study of Novel Implanted Vagus Nerve Stimulation Device to Treat Rheumatoid Arthritis," *Annals of the Rheumatic Diseases* 78, supplement 2 (June 2019): 264.

3. Peter Staats, Charles Emala, Bruce Simon, and J. P. Errico, "Neurostimulation for Asthma," in Elliot S. Krames, P. Hunter Peckham, and Ali R. Rezai, eds., *Neuromodulation: Comprehensive Textbook of Principles, Technologies, and Therapies* (Cambridge, MA: Academic Press, 2009), 1339–45.

4. Robert Kaczmarczyk, Dario Tejera, Bruce J. Simon, and Michael T. Heneka, "Microglia Modulation through External Vagus Nerve Stimulation in a Murine Model of Alzheimer's Disease," *Journal of Neurochemistry* 146 (December 2017): 76–85, doi:10.1111/jnc.14284.

5. Harold A. Sackeim et al., "Vagus Nerve Stimulation (VNS) for Treatment-Resistant Depression: Efficacy, Side Effects, and Predictors of Outcome," *Neuropsychopharmacology* 25, no. 5 (April 2001): 713–28, doi:10.1016/S0893-133X(01)00271-8.

6. Emily Battinelli Masi et al., "Identification of Hypoglycemia-Specific Neural Signals by Decoding Murine Vagus Nerve Activity," *Bioelectronic Medicine* 5 (2019): article 9, https://doi.org/10.1186/s42234-019-0025-z.

7. Sophie C. Payne, John B. Furness, Owen Burns, Alicia Sedo, Tomoko Hyakumura, Robert K. Shepherd, and James B. Fallon, "Anti-Inflammatory Effects of Abdominal Vagus Nerve Stimulation on Experimental Intestinal Inflammation," *Frontiers in Neuroscience* 13 (May 2019): 418, doi:10.3389/fnins.2019.00418.

8. Liping Zhou et al., "Low-Level Transcutaneous Vagus Nerve Stimulation Attenuates Cardiac Remodelling in a Rat Model of Heart Failure with Preserved Ejection Fraction," *Experimental Physiology* 104, no. 1 (January 2019): 28–38, https://doi.org/10.1113/EP087351.

9. Teresa H. Sanders et al., "Cognition-Enhancing Vagus Nerve Stimulation Alters the Epigenetic Landscape," *Journal of Neuroscience* 39 (February 2019): 3454–69.

10. Lynne Peeples, "Core Concept: The Rise of Bioelectric Medicine Sparks Interest among Researchers, Patients, and Industry," *Proceedings of the National Academy of Sciences of the United States of America* 116, no. 49 (December 2019): 24379–82, doi:10.1073/pnas.1919040116.

11. Ibid.

12. Sigrid Breit, Aleksandra Kupferberg, Gerhard Rogler, and Gregor Hasler, "Vagus Nerve as Modulator of the Brain-Gut Axis in Psychiatric and Inflammatory Disorders," *Frontiers in Psychiatry* 9 (March 2018): 44, doi:10.3389/fpsyt.2018.00044.

13. Bern Dibner, "Luigi Galvani," *Encyclopedia Britannica*, https://www.britannica.com/biography/Luigi-Galvani.

14. Alice Park, "Why It's Time to Take Electrified Medicine Seriously," *Time*, October 24, 2019, https://time.com/5709245/bioelectronic-medicine-treatments/.

15. Peeples, "Core Concept."

16. "Auricular Neurostimulation," Children's Hospital Wisconsin, https://chw.org/medical-care/gastroenterology-liver-and-nutrition-program/tests-and-treatments/auricular-neuro stimulation.

17. "FDA Permits Marketing of First Medical Device for Relief of Pain Associated with Irritable Bowel Syndrome in Patients 11–18 Years of Age," US Food & Drug Administration, June 7, 2019, https://www.fda.gov/news-events/press-announcements/fda-permits-marketing-first-medical-device-relief-pain-associated-irritable-bowel-syndrome-patients.

18. Robert M. G. Reinhart and John A. Nguyen, "Working Memory Revived in Older Adults by Synchronizing Rhythmic Brain Circuits," *Nature Neuroscience* 22 (2019): 820–27, https://doi.org/10.1038/s41593-019-0371-x.

19. Anne Trafton, "Ingestible Capsule Can Be Controlled Wirelessly," MIT News, December 13, 2018, http://news.mit.edu/2018/ingestible-pill-controlled-wirelessly-bluetooth-1213.

20. Jahyun Koo et al., "Wireless Bioresorbable Electronic System Enables Sustained Non-pharmacological Neuroregenerative Therapy," *Nature Medicine* 24 (2018): 1830–36, https://doi.org/10.1038/s41591-018-0196-2.

21. Park, "Why It's Time to Take Electrified Medicine Seriously."

22. Jamie Bell, "A Brief History of Brain Stimulation in Treating Mental Health Problems," NS Medical Devices, January 30, 2020, https://www.nsmedicaldevices.com/analysis/brain-stimula tion-therapy-history/.

23. Carnegie Mellon University, "Neuronal Targets to Restore Movement in Parkinson's Disease Model," ScienceDaily, May 8, 2017, https://www.sciencedaily.com/releases/2017/05/170508112420.htm.

24. "Optogenetics: Shedding Light on the Brain's Secrets," Scientifica, https://www.scientifica.uk.com/learning-zone/optogenetics-shedding-light-on-the-brains-secrets.

25. "Electroceuticals/Bioelectric Medicine Market Size, Share & Trends Analysis Report by Product (Implantable Cardioverter Defibrillators, Cardiac Pacemakers, Cochlear Implants), by Type, by Application, by End-Use, and Segment Forecasts 2019-2026," Research and Markets, September 2019, https://www.researchandmarkets.com/research/59wvsm/electroceuticalsb.

26. Nadia Ramlagan, "Monitoring Vital Signs through Stretchable Electronic 'Skin,'" American Association for the Advancement of Science, August 10, 2011, https://www.aaas.org/news/science-monitoring-vital-signs-through-stretchable-electronic-skin.

27. Yin Long et al., "Effective Wound Healing Enabled by Discrete Alternative Electric Fields from Wearable Nanogenerators," *ACS Nano* 12, no. 12 (2018): 12533–40, doi:10.1021/acsnano.8b07038, https://pubs.acs.org/action/showCitFormats?doi=10.1021%2Facsnano.8b07038&href=/doi/10.1021%2Facsnano.8b07038.

28. Charles Q. Choi, "These 3 Electroceuticals Could Help You Heal Faster," IEEE Spectrum, January 26, 2019, https://spectrum.ieee.org/semiconductors/devices/these-3-electroceuticals-could-help-you-heal-faster.

29. Sarah A. Stanley, Jeremy Sauer, Ravi S. Kane, Jonathan S. Dordick, and Jeffrey M. Friedman, "Remote Regulation of Glucose Homeostasis in Mice Using Genetically Encoded Nanoparticles," *Nature Medicine* 21 (2015): 92–98, https://doi.org/10.1038/nm.3730.

30. Jessica Timmons, "When Can a Fetus Hear?" Healthline, January 5, 2018, https://www.healthline.com/health/pregnancy/when-can-a-fetus-hear.

31. James L. Hallenbeck, "The Final 48 Hours," in *Palliative Care Perspectives* (New York: Oxford University Press, 2003), https://www.mywhatever.com/cifwriter/library/70/4998.html.

32. Lisa Trank, "Ancient Sound Technology: The Breath of Creation," Gaia, December 22, 2019, https://www.gaia.com/article/ancient-sound-technology-the-breath-of-creation.

33. "Talk: Pythagoras," Wikiquote, https://en.wikiquote.org/wiki/Talk:Pythagoras.

34. Trank, "Ancient Sound Technology."

35. Junling Gao, Hang Kin Leung, Bonnie Wai Yan Wu, Stavros Skouras, and Hin Hung Sik, "The Neurophysiological Correlates of Religious Chanting," *Scientific Reports* 9 (2019): article 4262, https://doi.org/10.1038/s41598-019-40200-w.

36. Gennady G. Knyazev, "EEG Delta Oscillations as a Correlate of Basic Homeostatic and Motivational Processes," *Neuroscience & Biobehavioral Reviews* 36, no. 1 (January 2012): 677–95, https://doi.org/10.1016/j.neubiorev.2011.10.002.

37. American Polarity Therapy Association, https://polaritytherapy.org/.

38. William Graves, "Using a Tuning Fork in Medical Situations," Fabrication Enterprises Inc., August 14, 2018, https://www.fab-ent.com/using-tuning-fork-medical-situations/.

39. John Davis, "Can a Tuning Fork Detect a Stress Fracture?" Runners Connect, https://runnersconnect.net/tuning-fork-stress-fracture/.

40. XMLA, "Nitric Oxide and Tuning Forks," Biosonics, July 13, 2017, https://biosonics.com/2017/07/13/nitric-oxide-tuning-forks/.

41. "Welcome to Biosonics," Biosonics, https://biosonics.com/about-us/.

42. Donald L. Schomer, MD, and Fernando H. Lopes da Silva, MD, PhD, eds., *Niedermeyer's Electroencephalography: Basic Principles, Clinical Applications, and Related Fields* (New York: Oxford University Press, 2018).

43. Robert F. Burkard, Manuel Don, and Jos J. Eggermont, *Auditory Evoked Potentials: Basic Principles and Clinical Application* (Philadelphia: Lippincott Williams & Wilkins, 2007).

44. Amitha Kalaichandran, "How Sound Baths Ended Up Everywhere," *New York Times*, August 3, 2019, https://www.nytimes.com/2019/08/03/style/self-care/sound-baths.html.

45. Corinne Purtill, "Turns Out 'Sound Healing' Can Be Actually, Well, Healing," Quartz, January 20, 2016, https://qz.com/595315/turns-out-sound-healing-can-be-actually-well-healing/.

46. Integratron, https://www.integratron.com/.

47. Sam V. Norman-Haignere, Nancy Kanwisher, Josh H. McDermott, and Bevil R. Conway, "Divergence in the Functional Organization of Human and Macaque Auditory Cortex Revealed by fMRI Responses to Harmonic Tones," *Nature Neuroscience* 22 (2019): 1057–60, https://doi.org/10.1038/s41593-019-0410-7.

48. Beth McGroarty, "Wellness Music," Global Wellness Summit, https://www.globalwellnesssummit.com/2020-global-wellness-trends/wellness-music/.

49. Zoe Cormier, "Music Therapy: The Power of Music for Health," *Science Focus*, October 9, 2018, https://www.sciencefocus.com/the-human-body/the-power-of-music-for-health/.

50. Pat Vaughan Tremmel, "Taking Music Seriously," Northwestern University News, July 20, 2010, https://www.northwestern.edu/newscenter/stories/2010/07/music-training-boosts-learning.html.

51. Marisa Aveling, "The Doctor Is In (Your Pocket): How Apps Are Harnessing Music's Healing Powers," *Pitchfork*, November 14, 2016, https://pitchfork.com/features/article/9976-the-doctor-is-in-your-pocket-how-apps-are-harnessing-musics-healing-powers/.

52. "Oliver Sacks, MD," https://www.oliversacks.com/inspired-by-sacks/.

53. Cormier, "Music Therapy."

54. Katie Moisse, Bob Woodruff, James Hill, and Lana Zak, "Gabby Giffords: Finding Words through Song," ABC News, November 8, 2011, https://abcnews.go.com/Health/w_MindBody News/gabby-giffords-finding-voice-music-therapy/story?id=14903987.

55. Simone Dalla Bella et al., "Gait Improvement via Rhythmic Stimulation in Parkinson's Disease Is Linked to Rhythmic Skills," *Scientific Reports* 7 (January 2017): 42005, doi:10.1038/srep42005.

56. Dalvin Brown, "Burnout Is Officially a Medical Condition, According to the World Health Organization," *USA Today*, May 28, 2019, https://www.usatoday.com/story/money/2019/05/28/burnout-official-medical-diagnosis-says-who/1256229001/.

57. Department of Economic and Social Affairs, Population Division, "World Population Ageing 2019," United Nations, https://www.un.org/en/development/desa/population/publications/pdf/ageing/WorldPopulationAgeing2019-Highlights.pdf.

58. Division of Population Health, "Mental Health and Aging in America (MAHA), Issue

Brief 1: What Do the Data Tell Us?" Centers for Disease Control and Prevention, 2008, http://www.cdc.gov/aging/pdf/mental_health.pdf.

59. "Fact Sheet: Chronic Disease Self-Management," National Council on Aging, 2016, https://www.ncoa.org/resources/fact-sheet-cdsmp/.

60. Daniel Leubner and Thilo Hinterberger, "Reviewing the Effectiveness of Music Interventions in Treating Depression," *Frontiers in Psychology* 8 (2017): 1109, doi:10.3389/fpsyg.2017.01109.

61. Jenny Quach and Jung-Ah Lee, "Do Music Therapies Reduce Depressive Symptoms and Improve QOL in Older Adults with Chronic Disease?" *Nursing* 47, no. 6 (June 2017): 58–63, doi:10.1097/01.NURSE.0000513604.41152.0c.

62. Myriam V. Thoma et al., "The Effect of Music on the Human Stress Response," *PLoS One* 8, no. 8 (2013): e70156, doi:10.1371/journal.pone.0070156.

63. Alexandra Linnemann, Beate Ditzen, Jana Strahler, Johanna M. Doerr, and Urs Nater, "Music Listening as a Means of Stress Reduction in Daily Life," *Psychoneuroendocrinology* 60 (June 2015): 82–90, doi:10.1016/j.psyneuen.2015.06.008.

64. Erin Kelly, "The Science behind Music: 5 Weird Brain Reactions Explained," All That's Interesting, June 26, 2017, https://allthatsinteresting.com/science-behind-music.

65. "Bob Marley Quotes," Goodreads, https://www.goodreads.com/quotes/4401-one-good-thing-about-music-when-it-hits-you-you.

66. McGill University, "Musical Chills: Why They Give Us Thrills," ScienceDaily, January 12, 2011, https://www.sciencedaily.com/releases/2011/01/110112111117.htm.

67. Peter R. Chai et al., "Music as an Adjunct to Opioid-Based Analgesia," *Journal of Medical Toxicology* 13, no. 3 (September 2017): 249–54, https://dx.doi.org/10.1007%2Fs13181-017-0621-9.

68. Recovery Unplugged, https://www.recoveryunplugged.com/.

69. Mike Iorfino, "Music Can Be a Viable Alternative to Medications in Reducing Anxiety before Anesthesia Procedures," Penn Medicine News, July 19, 2019, https://www.pennmedicine.org/news/news-releases/2019/july/music-can-be-a-viable-alternative-to-medications-in-reducing-anxiety-before-anesthesia-procedures.

70. Esther Inglis-Arkell, "How a Scientist's Least Important Discovery Became His Most Famous," Gizmodo, November 16, 2015, https://gizmodo.com/how-a-scientists-least-important-discovery-became-his-m-1742623371.

71. Gerald Oster, "Auditory Beats in the Brain," *Scientific American*, October 1973.

72. Rene Pierre Le Scouarnec et al., "Use of Binaural Beat Tapes for Treatment of Anxiety: A Pilot Study of Tape Preference and Outcomes," *Alternative Therapies in Health and Medicine* 7, no. 1 (February 2001): 58–63.

73. R. Padmanabhan, A. J. Hildreth, and D. Laws, "A Prospective, Randomised, Controlled Study Examining Binaural Beat Audio and PreOperative Anxiety in Patients Undergoing General Anaesthesia for Day Case Surgery," *Anaesthesia* 60, no. 9 (September 2005): 874–77, doi:10.1111/j.1365-2044.2005.04287.x.

74. Zeev N. Kain and Inna Maranets, "Preoperative Anxiety and Intraoperative Anesthetic Requirements," *Anesthesia & Analgesia* 91, no. 1 (July 2000): 1346–51.

75. Zeev N. Kain, Ferne Sevarino, Gerianne M. Alexander, Sharon Pincus, and Linda C. Mayes, "Preoperative Anxiety and Postoperative Pain in Women Undergoing Hysterectomy: A Repeated-Measures Design," *Journal of Psychosomatic Research* 49, no. 6 (December 2000): 417–22.

76. "Biography," Scientific Sounds, https://www.scientificsounds.com/dr-thompson/biography.

77. Lori Smith, "What Are Binaural Beats, and How Do They Work?" Medical News Today, September 30, 2019, https://www.medicalnewstoday.com/articles/320019.

78. Miguel Garcia-Argibay, Miguel A. Santed, and José M. Reales, "Efficacy of Binaural Auditory Beats in Cognition, Anxiety, and Pain Perception: A Meta-Analysis," *Psychological Research* 83, no. 2 (March 2019): 357–72, doi:10.1007/s00426-018-1066-8.

79. Nantawachara Jirakittayakorn and Yodchanan Wongsawat, "Brain Responses to a 6-Hz Binaural Beat: Effects on General Theta Rhythm and Frontal Midline Theta Activity," *Frontiers in Neuroscience* 11 (June 2017): 365, doi:10.3389/fnins.2017.00365.

80. Jennifer J. Newson and Tara C. Thiagarajan, "EEG Frequency Bands in Psychiatric Disorders: A Review of Resting State Studies," *Frontiers in Human Neuroscience* 12 (January 2019): 521, doi: 10.3389/fnhum.2018.00521.

81. Scientific Sounds, https://www.scientificsounds.com/.

82. Minji Lee, Chae-Bin Song, Gi-Hwan Shin, and Seong-Whan Lee, "Possible Effect of Binaural Beat Combined with Autonomous Sensory Meridian Response for Inducing Sleep," *Frontiers in Human Neuroscience* 13 (December 2019): 425, doi:10.3389/fnhum.2019.00425.

83. Emma L. Barratt and Nick J. Davis, "Autonomous Sensory Meridian Response (ASMR): A Flow-Like Mental State," *PeerJ* 3 (March 2015): e851, https://doi.org/10.7717/peerj.851.

84. Meghan Neal, "The Many Colors of Sound," *Atlantic*, February 16, 2016, https://www.theatlantic.com/science/archive/2016/02/white-noise-sound-colors/462972/.

85. Junhong Zhou, Dongdong Liu, Xin Li, Jing Ma, Jue Zhang, and Jing Fang, "Pink Noise: Effect on Complexity Synchronization of Brain Activity and Sleep Consolidation," *Journal of Theoretical Biology* 306 (August 2012): 68–72, doi:10.1016/j.jtbi.2012.04.006.

86. Nelly A. Papalambros et al., "Acoustic Enhancement of Sleep Slow Oscillations and Concomitant Memory Improvement in Older Adults," *Frontiers in Human Neuroscience* 11 (March 2017): 109, doi:10.3389/fnhum.2017.00109.

87. Kirsten Nunez, "What Is Pink Noise and How Does It Compare with Other Sonic Hues?" Healthline, June 21, 2019, https://www.healthline.com/health/pink-noise-sleep.

88. Ibid.

89. Neal, "The Many Colors of Sound."

90. Tim Watson, "Ultrasound in Contemporary Physiotherapy Practice," *Ultrasonics* 48, no. 4 (August 2008): 321–29, doi:10.1016/j.ultras.2008.02.004.

91. Fareeha Amjad, Hassan Anjum Shahid, Sana Batool, Ashfaq Ahmad, and Imran Ahmed, "A Comparison on Efficacy of Transcutaneous Electrical Nerve Stimulation and Therapeutic Ultrasound in Treatment of Myofascial Trigger Points," *Khyber Medical University Journal* 8, no. 1 (2016): 3–6, https://www.kmuj.kmu.edu.pk/article/view/15769.

05: FREE ENERGY

1. Meg Neal, "The Eternal Quest for Aether, the Cosmic Stuff That Never Was," *Popular Mechanics*, October 19, 2018, https://www.popularmechanics.com/science/energy/a23895030/aether/.

2. "About," ARPA-E, US Department of Energy, https://arpa-e.energy.gov/?q=arpa-e-site-page/about.

3. "Projects," ARPA-E, US Department of Energy, https://arpa-e.energy.gov/?q=arpa-e-site-page/projects.

4. Matthew Hole, "We Won't Have Fusion Generators in 5 Years. But the Holy Grail of Clean Energy May Still Be on Its Way," SingularityHub, March 5, 2020, https://singularityhub.com/2020/03/05/we-wont-have-fusion-generators-in-5-years-but-the-holy-grail-of-clean-energy-may-still-be-on-its-way/.

5. "What Is ITER?" ITER Organization, https://www.iter.org/proj/inafewlines.

6. Craig Busse, "What Is the Difference between Nuclear Fusion and Cold Fusion?" PhysLink.com, https://www.physlink.com/education/askexperts/ae330.cfm.

7. Colin Schultz, "The Man Who 'Discovered' Cold Fusion Just Passed Away," *Smithsonian Magazine*, August 8, 2012, https://www.smithsonianmag.com/smart-news/the-man-who-discovered-cold-fusion-just-passed-away-15895414/.

8. Ibid.

9. Jonathan Tennenbaum, "Cold Fusion: A Potential Energy Gamechanger," *Asia Times*, November 14, 2019, https://asiatimes.com/2019/11/cold-fusion-1-a-potential-energy-gamechanger/.

10. Ibid.
11. Ilker Koksal, "A Massive Investment: Google Announces 18 New Renewable Energy Deals," *Forbes*, October 2, 2019, https://www.forbes.com/sites/ilkerkoksal/2019/10/02/a-massive-investment-google-announces-18-new-renewable-energy-deals/#6602105c5024.
12. Tennenbaum, "Cold Fusion."
13. Koksal, "A Massive Investment."
14. "ALPHA," ARPA-E, US Department of Energy, https://arpa-e.energy.gov/?q=arpa-e-programs/alpha.
15. "Climate Change Indicators: Weather and Climate," EPA, https://www.epa.gov/climate-indicators/weather-climate.
16. Neel V. Patel, "WTF Is Zero Point Energy and How Could It Change the World?" Inverse, August 4, 2017, https://www.inverse.com/article/35077-wtf-is-zero-point-energy.
17. Stephen Reucroft and John Swain, "What Is the Casimir Effect?" *Scientific American*, June 22, 1998, https://www.scientificamerican.com/article/what-is-the-casimir-effec/.
18. Bernhard Haisch, Alfonso Rueda, and H. E. Puthoff, "Advances in the Proposed Electromagnetic Zero-Point Field Theory of Inertia," 34th AIAA/ASME/SAE/ASEE AIAA Joint Propulsion Conference, AIAA paper 98-3143, 1998.
19. Jayshree Pandya, "A Quantum Revolution Is Coming," *Forbes*, May 25, 2019, https://www.forbes.com/sites/cognitiveworld/2019/05/25/a-quantum-revolution-is-coming/#50f21014374d.
20. Institute for Advanced Studies at Austin, EarthTech International, Inc., https://earthtech.org/.
21. Adam Hadhazy, "Will Space-Based Solar Power Finally See the Light of Day?" *Scientific American*, April 16, 2009, https://www.scientificamerican.com/article/will-space-based-solar-power-finally-see-the-light-of-day/.
22. Joe Pinkstone, "China Wants to Build a Power Station in SPACE by 2030 That Will Beam Solar Energy to Earth while in Orbit," *Daily Mail*, February 18, 2019, https://www.dailymail.co.uk/sciencetech/article-6717393/China-wants-build-solar-power-station-SPACE.html.

06: THE MIND FIELD

1. Joseph Banks Rhine and J. Gaither Pratt, *Extra Sensory Perception After Sixty Years: A Critical Appraisal of the Research in Extra Sensory Perception* (New York: Henry Holt, 1940).
2. "Dean Radin Quotes," GoodReads, https://www.goodreads.com/author/quotes/2925230.Dean_Radin.
3. "JREF Challenge FAQ," James Randi Educational Foundation, https://web.randi.org.
4. Richard Tarnas, *Cosmos and Psyche: Intimations of a New World View* (New York: Viking, 2006).
5. Emile Deschamps, *Oeuvres completes: Tomes I-VI*, Reimpr. de l'ed. de Paris 1872-74.
6. SQuire Rushnell, *When God Winks* (New York: Atria, 2006).
7. Erik Davis, *High Weirdness: Drugs, Esoterica, and Visionary Experience in the Seventies* (Cambridge, MA: MIT Press, 2019), 147–48.
8. F. David Peat, *Synchronicity: The Bridge between Matter and Mind* (New York: Bantam, 1987).
9. Tam Hunt and Jonathan W. Schooler, "The Easy Part of the Hard Problem: A Resonance Theory of Consciousness," *Frontiers in Human Neuroscience* 13 (October 2019): 378, doi:10.3389/fnhum.2019.00378.
10. Tam Hunt, "The Hippies Were Right: It's All about Vibrations, Man!" *Scientific American*, December 5, 2018, https://blogs.scientificamerican.com/observations/the-hippies-were-right-its-all-about-vibrations-man/.
11. Eric Schwitzgebel, "The Crazyist Metaphysics of Mind," July 12, 2013, https://faculty.ucr.edu/~eschwitz/SchwitzPapers/CrazyMind-130712.pdf.
12. Tam Hunt, "Could Consciousness All Come Down to the Way Things Vibrate?" The Conversation, November 9, 2018, https://theconversation.com/could-consciousness-all-come-down-to-the-way-things-vibrate-103070.

13. Rupert Sheldrake, *The Sense of Being Stared At: And Other Aspects of the Extended Mind* (New York: Harmony, 2003)

14. Lianpeng Z. Song, Gary E. Schwartz, and Linda G. S. Russek, "Heart-Focused Attention and Heart-Brain Synchronization: Energetic and Physiological Mechanisms," *Alternative Therapies in Health and Medicine* 4, no. 5 (1998): 44–62.

15. HeartMath Institute, *Science of the Heart*, chapter 6: Energetic Communication.

16. Rollin McCraty, Annette Deyhle, and Doc Childre, "The Global Coherence Initiative: Creating a Coherent Planetary Standing Wave," *Global Advances in Health and Medicine* 1, no. 1 (March 2012): 64–77, doi:10.7453/gahmj.2012.1.1.013.

17. Association for Psychological Science, "Sleep Makes Your Memories Stronger, and Helps with Creativity," ScienceDaily, December 17, 2010, www.sciencedaily.com/releases/2010/11/101113165441.htm.

18. Michelle Neider, Edward F. Pace-Schott, Erica Forselius, Brian Pittman, and Peter T. Morgan, "Lucid Dreaming and Ventromedial versus Dorsolateral Prefrontal Task Performance," *Consciousness and Cognition* 20, no. 2 (June 2011): 234–44.

19. NextMind, https://www.next-mind.com/.

20. Ursula Voss et al., "Induction of Self-Awareness in Dreams through Frontal Low Current Stimulation of Gamma Activity," *Nature Neuroscience* 17 (2014): 810–12, https://doi.org/10.1038/nn.3719.

21. Ibid.

22. Antoine Lutz, Lawrence L. Greischar, Nancy B. Rawlings, Matthieu Ricard, and Richard J. Davidson, "Long-Term Meditators Self-Induce High-Amplitude Gamma Synchrony during Mental Practice," *Proceedings of the National Academy of Sciences of the United States of America* 101, no. 46 (November 2004): 16369–73, doi:10.1073/pnas.0407401101.

23. Joe Nickell, "Remotely Viewed? The Charlie Jordan Case," *Skeptical Inquirer*, March 1, 2001, https://skepticalinquirer.org/newsletter/remotely-viewed-the-charlie-jordan-case/.

24. Russell Targ, *The Reality of ESP: A Physicist's Proof of Psychic Abilities* (Wheaton, IL: Quest Books, 2012), 4, 14, 23.

25. Michael D. Mumford, Andrew M. Rose, and David A. Goslin, "An Evaluation of Remote Viewing: Research and Applications," American Institutes for Research, Washington, DC, September 29, 1995, https://www.cia.gov/library/readingroom/docs/CIA-RDP96-00791R000200180006-4.pdf.

26. Alex Heard, "Close Your Eyes and Remote View This Review," *San Diego Union-Tribune*, April 10, 2010: "This so-called remote viewing operation continued for years, and came to be known as Star Gate."

27. "What Is Remote Viewing?" Farsight Institute, https://farsight.org/WhatIsRemoteViewing.html.

28. Michael A. Persinger, "The Neuropsychiatry of Paranormal Experiences," *Neuropsychiatric Practice and Opinion* 13, no. 4 (November 2001): 515–24, doi:10.1176/jnp.13.4.515.

29. Andrea Michelson, "Five Scientific Explanations for Spooky Sensations," *Smithsonian Magazine*, October 20, 2019, https://www.smithsonianmag.com/smart-news/five-scientific-explanations-spooky-sensations-180973436/.

30. David Biello, "Searching for God in the Brain," *Scientific American*, October 2007, https://www.scientificamerican.com/article/searching-for-god-in-the-brain/.

31. Andrew Newberg, "Research Questions: How Do Meditation and Prayer Change Our Brains?" http://www.andrewnewberg.com/research.

32. Andrew B. Newberg, *Principles of Neurotheology* (Burlington, VT: Ashgate Publishing, 2010).

33. David Bryce Yaden et al., "The Noetic Quality: A Multimethod Exploratory Study," *Psychology of Consciousness: Theory, Research, and Practice* 4, no. 1 (March 2017): 54–62, https://doi.org/10.1037/cns0000098.

34. "Our Origins: A Shift in Perspective," Institute of Noetic Sciences, https://noetic.org/about/origins/.

35. Duke University Medical Center, "Prayer, Noetic Studies Feasible; Results Indicate Benefit to Heart Patients," EurekAlert, October 31, 2001, https://www.eurekalert.org/pub_releases/2001-10/dumc-pns102901.php.

36. Mitchell W. Krucoff et al., "Integrative Noetic Therapies as Adjuncts to Percutaneous Intervention during Unstable Coronary Syndromes: Monitoring and Actualization of Noetic Training (MANTRA) Feasibility Pilot," *American Heart Journal* 142, no. 5 (November 2001): 760–69, https://doi.org/10.1067/mhj.2001.119138.

37. Marilyn Schlitz and Tina Amorok with Marc S. Micozzi, *Consciousness & Healing: Integral Approaches to Mind-Body Medicine* (London: Churchill Livingstone, 2005).

38. "Special Care Focus: Rise of the Human Spirit," HeartMath Institute, https://www.heartmath.org/calendar-of-events/special-care-focus-2-rise-of-the-human-spirit/.

39. Lewis Rowell, *Music and Musical Thought in Early India* (Chicago: University of Chicago Press, 1992), 48.

40. Regal, *Pseudoscience*, 29: "Other than anecdotal eyewitness accounts, there is no evidence of the ability to astral project, the existence of other planes, or of the Akashic Record."

41. Rudolf Steiner, *The Submerged Continents of Atlantis and Lemuria, Their History and Civilization. Being Chapters from the Âkâshic Records* (London: Theosophical Publishing Society, 1911).

42. "Biography," http://ervinlaszlo.com/index.php/biography.

43. The Soul Rider, LLC, Leaders & Advisors, https://thesoulrider.net/.

44. Ibid.

07: ALICE IN FUTURELAND

1. Stephen Battersby, "Could Alien Life Exist in the Form of DNA-Shaped Dust?" New Scientist, August 10, 2007, https://www.newscientist.com/article/dn12466-could-alien-life-exist-in-the-form-of-dna-shaped-dust/.

Illustration Credits

Most of the photographs throughout this book were sourced from Pexels.com, a free stock photo and video website. Thank you to the photographers and individuals who support open-source photography organizations.

INTRODUCTION: GOOD VIBRATIONS

pp. x–xi (*clockwise from upper left*): City Street with Electric Billboards: Photo by Nextvoyage from Pexels; Elephant: Photo by Rudolf Kirchner from Pexels; Dandelion Seed: Photo by Taner Soyler from Pexels; Woman in Field: Photo by Alex Fu from Pexels, Haloed Rocks: Photo by Ian Beckley from Pexels; Eye: Photo by Adrian Swancar on Unsplash.

p. xii: Digital Framework: Photo by Pixaby from Pexels.

p. xiii: Glittered Face: Photo by 3Motional Studio from Pexels.

p. xv: Woman with Dark Hair: Photo by Run Away from Pexels.

p. xvi: Pipe Organ: Photo by Julian Larcher on Unsplash.

p. xviii: Hand Reaching: jessica-flavia-unsplash.

p. xix: Cymatics Book: © 2001 MACROmedia Publishing, Eliot, ME. Used by permission.

pp. xx–xxi: Woman with Closed Eyes: Photo by Chermiyi Mohamed from Pexels.

p. xxiii: Woman in Front of Sky: Photo by Nappy from Pexels.

CHAPTER 01: WE LIVE IN AN ELECTROMAGNETIC WORLD

p. xxiv: City Street with Electric Billboards: Photo by Nextvoyage from Pexels.

p. 3: Black Figure in Neon Hallway: Photo by Naveen Annam from Pexels.

p. 5: Woman in Front of Neon Bars: Photo by Mahdis Mousavi on Unsplash.

pp. 6-7: Planet Earth at Night: Photo by NASA on Unsplash.

p. 8: Neon Pink Fabric Waves: Photo by Louys from Pexels.

p. 11: Man in Blue Light: Photo by PIXABY from Pexels.

pp. 12-13: Woman under Electric Umbrella: Photo by Matheus Bertelli from Pexels.

p. 15: Woman in Front of Night Sky: Photo by luizclas from Pexels.

p. 16: Woman with Light Bulbs: Photo by Isadora Menezes from Pexels.

p. 20: Flowers on Person's Back: Life of Pex from Pexels.

p. 23: Woman and Man Doing Tai Chi: Photo by Craig Adderley from Pexels.

pp. 24-25: Daisies: Juan Salamanca from Pexels.

p. 30: Woman's Hand with Flowers: Paula Ropero from Pexels.

p. 35: Systems Change Map: Subtle Energy & Biofield Healing report, Consciousness and Healing Initiative.

p. 36: Woman and Man Facing Each Other: Photo by freestocks.org from Pexels.

p. 39: Woman and Man Facing Each Other in Blue Light: Photo by Masha Raymers from Pexels.

pp. 42-43: Hand Holding Button: Photo by Mark Tacatani from Pexels.

p. 44: Kids Chasing Bubbles: Photo by Máximo from Pexels.

p. 46: Family Laughing: Photo by Nappy from Pexels.

pp. 48-49: Face over Urban Landscape: Photo by Gift Habeshaw from Pexels.

p. 51: Woman's Feet on Leaves: Photo by Daria Shevtsova from Pexels.

p. 55: Man with Veil: Photo by Milena Santos from Pexels.

p. 56: Hand with Lens: Photo by Mauricio Mascaro from Pexels.

p. 61: Boy Looking at Blue Light: Photo by Chris LeBoutillier from Pexels.

p. 65: Hand Tossing Globe: Photo by Valentin Antonucci from Pexels.

CHAPTER 02: CATCHING NATURE'S VIBE

p. 66: Elephant: Photo by Rudolf Kirchner from Pexels.

p. 68: Trees Lining Road: Photo by Philipp Knape on Unsplash.

p. 70: Starling: Photo by flckr from Pexels.

p. 71: Hand Tossing Rock: Photo by Miguel Bruna on Unsplash.

CHAPTER 04: FREQUENCY HEALING

p. 158: Dandelion Seed: Photo by Taner Soyler from Pexels.
p. 161: Woman in White Dress: Photo by Nastyasensi from Pexels.
p. 162: Hand Holding Sparkler: Photo by Luke Barkey from Pexels.
p. 165: Lights Surrounding Spine: Photo by Venu Gopal from Pexels.
p. 165: Close-Up of Gold Prongs: Photo by Pixaby from Pexels.
p. 167: Woman with Blue Mask in Light Tunnel: Photo by Zichuan Han from Pexels.
p. 169: Woman Pointing to Temple: Photo by engin akyurt from Pexels.
pp. 170-71: Band of Gold Electrical Charges: Photo by Francesco Ungaro from Pexels.
p. 173: Woman Dancing under Lights: Photo by Craig Adderley from Pexels.
pp. 174-75: Woman Wearing VR Headset: Photo by Sound from Pexels.
p. 177: Woman Wearing White Headphones: Photo by Sound from Pexels.
p. 179: Gong and Mallets: Photo by FilmBros on Pexels.
pp. 180-81: Tibetan Singing Bowls: Photo by Singing bowls on Pexels.
pp. 182-83: Tuning Forks: Photo by Mariia Loginovskaia from Adobe Stock.
p. 186: Hand Playing Singing Bowl: Photo by Singing Bowls from Pexels.
p. 188: Woman Meditating in Forest: Photo by Oluremi Adebayo from Pexels.
pp. 190-91: Woman Wearing Earbuds: Photo by bruce mars from Pexel.
p. 192: Metronome: Photo by Pixaby from Pexel.
p. 195: Elderly Man Wearing Headphones: Photo by Andrea Piacquadio from Pexels.
pp. 196-97: Woman Playing Piano: Photo by Andrea Piacquadio from Pexels.
pp. 198-99: Concert-Goers: Photo by anna-mw.jpg from Pexels.
p. 200: Colored Vinyl Disc: Photo by Stas Knop from Pexels.
p. 201: Mushroom: Photo by Elina Krima from Pexels.
pp. 202-3: Woman with Fingers Pressed to Headphones: Photo by Andrea Piacquadio from Pexels.
p. 206: Man Stretching and Smiling: Photo by Andrea Placquadio from Pexels.
pp. 208-9: Back of Woman Wearing Headphones: Photo by Andrea Placquadio from Pexels.
p. 211: Woman at Desk Wearing Headphones: Photo by Andrea Placquadio from Pexels.
p. 215: Speech Icon: Photo by Miguel Á. Padriñán from Pexels.
p. 216: Woman in Profile: Photo by Prasanth Inturi from Pexels.

CHAPTER 05: FREE ENERGY

p. 218: Haloed Rocks: Photo by Ian Beckley from Pexels.
p. 220: Silhouette of Woman: Photo by luizclas from Pexels.
p. 222: Light Bulb over Black Background: Photo by Pixaby from Pexels.
p. 226: Sunset through Tree Silhouette: Photo by Pixaby from Pexels.
p. 228: Diagram and Mathematical Formulae: 1000pixels from Adobe Stock.
p. 231: Twilight Sky: Photo by Bra Kou from Pexels.
p. 233: Light Waves on Rooftop: Photo by Federico Beccari on Unsplash.
p. 234: Man in Light Wave Spiral: Photo by Ioannis Ritos from Pexels.
p. 236: Light Ring around Woman's Head: Photo by Dark Indigo from Pexels.
p. 239: Light Bulb in Dark Room: Photo by Rahul from Pexels.
p. 240: Hand Reaching for the Sun: Photo by Jonas Ferlin from Pexels.

CHAPTER 06: THE MIND FIELD

p. 242: Woman in Field: Photo by Alex Fu from Pexels.
p. 245: Woman Looking at Pictures on Wall: Photo by cottonbro from Pexels.
p. 246: Woman Holding Hologram over Face: Photo by Elina Krima from Pexels.
pp. 248-49: Woman Basking in Sun: Photo by Radu Florin from Pexels.
p. 251: Long Exposure of Night Sky: Photo by Jakub Novacek from Pexels.
p. 253: Silhouette of Girl in Field: Photo by luizclas from Pexels.
pp. 254-55: Gold Heart and Brain: Purchased from Adobe.
p. 257: Man Lying on Grass: Photo by Pixaby on Pexels.
pp. 258-59: Girl with Stars on Face: Photo by Allie Smith from Pexels.
p. 260: Silhouette of Woman in Lotus Position: Photo by Prasanth Inturi from Pexels.
pp. 262-63: View of Road: Photo by George Becker from Pexels.
p. 265: Blue Sky: Photo by Elia Clerici from Pexels.
pp. 266-67: Silhouette through Curtain: Photo by Steinar Engeland on Unsplash.

Index

About the Authors

SPUTNIK FUTURES is a research consultancy that specializes in frontier futures, long-range intelligence that enables organizations to resonate in a world of constant and dynamic change. Sputnik has a public archive of original video interviews with thought leaders around the world, from Nobel Prize winners to acclaimed innovators. Sputnik's cofounders have provided strategic foresight consultation to multinational corporations for more than twenty-five years. Alice in Futureland is Sputnik's media platform that inspires imagination and thought on the exciting possibilities at the intersection of art, science, technology, and culture.